# Student's Solutions Manual

to accompany

# Mathematics for Elementary Teachers
## *A Conceptual Approach*

Seventh Edition

**Albert B. Bennett, Jr.**
*Univerity of New Hampshire*

**L. Ted Nelson**
*Portland State University*

Prepared by
**Joseph Ediger**
*Portland State University*

Boston   Burr Ridge, IL   Dubuque, IA   New York   San Francisco   St. Louis
Bangkok   Bogotá   Caracas   Kuala Lumpur   Lisbon   London   Madrid   Mexico City
Milan   Montreal   New Delhi   Santiago   Seoul   Singapore   Sydney   Taipei   Toronto

The **McGraw·Hill** Companies

Student's Solutions Manual to accompany
MATHEMATICS FOR ELEMENTARY TEACHERS: A CONCEPTUAL APPROACH, SEVENTH EDITION
ALBERT B. BENNETT, JR. AND L. TED NELSON

Published by McGraw-Hill Higher Education, an imprint of The McGraw-Hill Companies, Inc., 1221 Avenue of the Americas, New York, NY 10020. Copyright © 2007 by The McGraw-Hill Companies, Inc. All rights reserved.

No part of this publication may be reproduced or distributed in any form or by any means, or stored in a database or retrieval system, without the prior written consent of The McGraw-Hill Companies, Inc., including, but not limited to, network or other electronic storage or transmission, or broadcast for distance learning.

1 2 3 4 5 6 7 8 9 0 CUS/CUS 0 9 8 7 6

ISBN-13: 978-0-07-321329-3
ISBN-10: 0-07-321329-2

www.mhhe.com

# To the Student:

New to this edition are the "*Writing and Discussion*" and "*Making Connections*" sections at the end of each set of exercises and problems. Solutions to these problems are not given in this manual. They are usually open-ended questions and do not have unique solutions. Many of them bring up the same kinds of issues that teachers face every day. Use these problems as an opportunity to put yourself into the role of a teacher. As you prepare to write responses, think about what you as a teacher will do to help the learning process of your students. It is also helpful to reflect on your own learning of mathematics at the same time.

Here are a few more suggestions for using this solutions manual to enhance your learning:

- Always work on the problem first before checking with the solutions manual. Usually some struggle is necessary for real learning to take place. By looking at a solution written by someone else before doing your own thinking you may lose an opportunity for learning.

- The solutions given here are usually not the only correct ways to work the problems. If you worked a problem in a different way and got a correct answer then your solution is probably just as good or better. On some problems more than one solution method has been shown.

- When reading a solution, be sure that each part makes sense to you. It may be helpful to rephrase things in your own words.

- Additional questions or comments have been added at the end of some of the solutions. [*These are written in italics and placed in brackets like this.*] Sometimes these present opportunities for you to challenge yourself and extend your learning.

- Look for the fun in learning mathematics. If you find yourself really interested in a particular problem, see whether other related questions or patterns occur to you. Let your interests guide your learning.

# Acknowledgments:

Many thanks to Al Bennett and Ted Nelson for their support and encouragement in this project. Thanks also to Amy Gembala of McGraw-Hill and to the faculty and staff of the Portland State University Department of Mathematics and Statistics. Special appreciation goes to my life-partner Kathryn Falkenstern for her constant love and support.     JRE

# CONTENTS

## 1 Problem Solving

| | | |
|---|---|---|
| 1.1 | Introduction to Problem Solving | 1 |
| 1.2 | Patterns and Problem Solving | 7 |
| 1.3 | Problem Solving with Algebra | 13 |
| | *Chapter 1 Test* | 21 |

## 2 Sets, Functions, and Reasoning

| | | |
|---|---|---|
| 2.1 | Sets and Venn Diagrams | 26 |
| 2.2 | Functions and Graphs | 30 |
| 2.3 | Introduction to Deductive Reasoning | 37 |
| | *Chapter 2 Test* | 42 |

## 3 Whole Numbers

| | | |
|---|---|---|
| 3.1 | Numeration Systems | 47 |
| 3.2 | Addition and Subtraction | 52 |
| 3.3 | Multiplication | 58 |
| 3.4 | Division and Exponents | 65 |
| | *Chapter 3 Test* | 71 |

## 4 Number Theory

| | | |
|---|---|---|
| 4.1 | Factors and Multiples | 76 |
| 4.2 | Greatest Common Divisor and Least Common Multiple | 82 |
| | *Chapter 4 Test* | 87 |

## 5 Integers and Fractions

| | | |
|---|---|---|
| 5.1 | Integers | 91 |
| 5.2 | Introduction to Fractions | 96 |
| 5.3 | Operations with Fractions | 100 |
| | *Chapter 5 Test* | 107 |

## 6 Decimals: Rational and Irrational Numbers

| | | |
|---|---|---|
| 6.1 | Decimals and Rational Numbers | 112 |
| 6.2 | Operations with Decimals | 117 |

| | | |
|---|---|---|
| 6.3 | *Ratio, Percent, and Scientific Notation* | 122 |
| 6.4 | *Irrational and Real Numbers* | 127 |
| *Chapter 6 Test* | | 132 |

## 7 Statistics

| | | |
|---|---|---|
| 7.1 | *Collecting and Graphing Data* | 136 |
| 7.2 | *Describing and Analyzing Data* | 143 |
| 7.3 | *Sampling, Predictions, and Simulations* | 150 |
| *Chapter 7 Test* | | 154 |

## 8 Probability

| | | |
|---|---|---|
| 8.1 | *Single-Stage Experiments* | 158 |
| 8.2 | *Multi-Stage Experiments* | 163 |
| *Chapter 8 Test* | | 170 |

## 9 Geometric Figures

| | | |
|---|---|---|
| 9.1 | *Plane Figures* | 174 |
| 9.2 | *Polygons and Tessellations* | 178 |
| 9.3 | *Space Figures* | 182 |
| 9.4 | *Symmetric Figures* | 187 |
| *Chapter 9 Test* | | 191 |

## 10 Measurement

| | | |
|---|---|---|
| 10.1 | *Systems of Measurement* | 196 |
| 10.2 | *Area and Perimeter* | 199 |
| 10.3 | *Volume and Surface Area* | 204 |
| *Chapter 10 Test* | | 209 |

## 11 Motions in Geometry

| | | |
|---|---|---|
| 11.1 | *Congruence and Constructions* | 213 |
| 11.2 | *Congruence Mappings* | 217 |
| 11.3 | *Similarity Mappings* | 223 |
| *Chapter 11 Test* | | 227 |

# Chapter 1 Problem Solving
## Section 1.1

1. During the first 24 hours the snail climbs up 4 feet and slips back down 2 feet, so it reaches a maximum height of 4 feet and there is a net gain of 2 feet. It seems as if the snail should take 10 days to get out because it makes a net gain of 2 feet each day. This is not correct because on the 8$^{th}$ day it ends up at 16 feet and then on the 9$^{th}$ day it climbs up the remaining 4 feet to get out. It doesn't slide back down on the 9$^{th}$ day because it is already out of the well. The figure below shows one possible solution that uses a drawing.

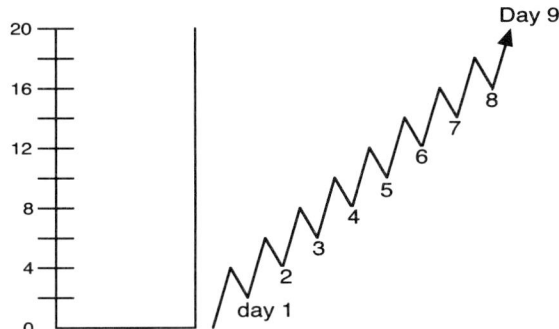

3. Here is a sketch of the two ropes laid end to end next to each other. The longer rope has an extra 26 feet in length, so if the 26 feet is marked off separately on the longer rope then the remaining section is the same length as the first rope. Together this is 130 – 26 = 104 feet. So the shorter rope is 104/2 = 52 feet long and the longer one is 52 + 26 = 78 feet long.

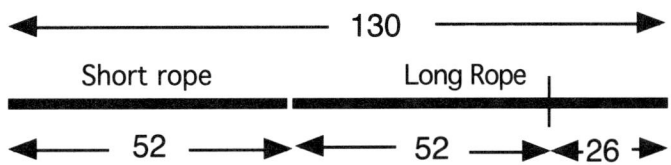

5. a. Under the old plan writing 10 checks will cost $2 plus 10 times .15, or 2 + 1.50 = $3.50. Under the new plan it is $3 + (10 × .08) = 3.00 + .80 = $3.80.

b,c. Here is a partial table for this problem. Note that for a small amount of checks the old plan is cheaper and as the number of checks increases, the cost for the old plan goes up faster than the cost for the new plan. At 15 checks the new plan becomes cheaper than the old plan.

| Checks | Old Plan | New Plan |
|---|---|---|
| 12 | $3.80 | $3.96 |
| 13 | $3.95 | $4.04 |
| 14 | $4.10 | $4.12 |
| 15 | $4.25 | $4.20 |
| 17 | $4.55 | $4.36 |
| 19 | $4.85 | $4.52 |

d. Each time the number of checks increases by one, the new plan gains 15 – 8 = 7 cents in value against the old plan. For example, in the table above we can see that at 12 checks the new plan is .16 more than the old, but at 13 checks it is only .09 higher. By the time 19 checks have been written, the new plan is 33 cents cheaper than the old one.

2     Chapter 1   Problem Solving

7.  One approach to this problem is to create a table similar to the one below. Another good way to create a table for this problem would be to include two additional columns, one for the cost of the postcards and one for the cost of the letters. This table was started with a guess at the solution of 10 postcards and 5 letters. If one started by guessing 3 postcards and 12 letters, then the cost would be too high and the number of letters would need to be reduced. Note that the total of the number of postcards and number of letters must always be 15.

| No. of Postcards | No. of Letters | Cost |
| --- | --- | --- |
| 10 | 5 | 2.00 + 1.60 = $3.60 |
| 9 | 6 | 1.80 + 1.92 = $3.72 |
| 7 | 8 | 1.40 + 2.56 = $3.96 |
| 6 | 9 | 1.20 + 2.88 = $4.08 |

An alternative approach to this problem would be to notice that each time there is one additional letter and one less postcard the cost increases by 12 cents. This approach could be used in conjunction with the table or in a solution that did not use a table.

9.  a,b,c.
    The most difficult part of this problem may be in reading and understanding. The sum of the digits of the two-digit number 29 is 2 + 9 = 11. We are looking for pairs of two-digit numbers which are different, but both have the same two digits. This means they will be numbers with their digits reversed, such as 29 and 92, 17 and 71, 53 and 35, etc. But we only want numbers whose digits add to 10. So the only possible pairs of two-digit numbers that could work are 19 and 91, 28 and 82, 37 and 73, 46 and 64, and maybe 55 and 55. We want the pair with a difference of 54, so it must be 28 and 82.

    d.  Pairs of two-digit numbers with sums of 12 are: 39 and 93, 48 and 84, 57 and 75, 66 and 66. The difference between 39 and 93 is 54.

11. a.  The sketch shows that we start with the 9 gallon container filled and the 4 gallon container empty. Then the 9 gallon container is dumped into the 4 gallon container twice, so that exactly one gallon is left in the 9 gallon container.

    b.  To measure 6 gallons, use the procedure from part a to measure one gallon into the 9 gallon container. Then dump this one gallon into the 4 gallon container. Next fill the 9 gallon container and pour this into the 4 gallon container until it is just full. Since it already had 1 gallon in it, 3 more gallons will be poured out, leaving 6 gallons in the large container.

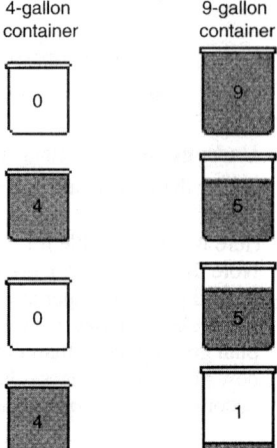

## 1.1 Introduction to Problem Solving

13. a. The yellow tiles are touching at the corners, and the directions say that no tile should touch the same color at any point.

    b. The center tile touches all of the other tiles, so if it were the same color as any of them it would violate that condition.

    c. With a blue tile in the center and reds in the corners, pairs of the four remaining openings touch each other at their corners. So one additional color will not work. The square can be formed with four colors, but not with three. One reason three will not work is that there are several places in the 3 × 3 square where four different squares meet in a single point. This forces at least four different colors.

    d. If the same color can touch at a corner, but not on an edge, then a checkerboard pattern will work. Two colors are sufficient for this problem.

15. a. The single square on top will end up opposite the square marked Bottom, so it will be the Top. In the row of four squares in the middle, Front and Back need to alternate with Left and Right. So the second square in this row needs to be labeled Right. Imagine the cube formed, with this square on the Right, then the first square on the left will be facing you and should be labeled Front. That leaves the third square for the Back.

    b. Working from right to left in the middle row, the square to the left of Bottom should be Front, since it will be opposite Back. Next to Front will be Top, since it is opposite Bottom. Now imagine folding the pieces up and placing it with Bottom down, etc. The square above Bottom must be Right and the square below Top must be Left.
    *[Note: If these problems are difficult to visualize, you might try building the model.]*

17. a. Girl A needs to give 30 chips to girl B and 20 to girl C, so she must give up 50 of her 70 chips. At the end of this round, girl A has 20 chips, girl B has 60, and girl C has 40.

    b. We will construct the scores at the end of the second round by working backward from the scores at the end of the third round, assuming that girl C lost the third round. Since both girl A and B doubled their scores to 40 in the third round, they must have had 20 each at the end of the second round. Girl C had to give them each 20 of her chips, so she must have had 40 + 20 + 20 = 80 chips at the end of the second round.
    A had 20; B had 20; C had 80.

    c. Continuing to work backward, if girl B lost the second round, then she doubled A's score from 10 to 20 and doubled C's from 40 to 80. So B gave away 10 + 40 = 50 chips in the second round. So the score at the end of the first round was: A had 10; B had 70; C had 40. If girl A lost the first round, she gave 35 chips to B and 20 chips to C. So girl A started the game with 10 + 35 + 20 = 65 chips. B started with 70 − 35 = 35 chips and C started with 20 chips. The original distribution was 65 for A, 35 for B, and 20 for C.

    d. Suppose girl C had lost the first round. She would need to give 65 chips to A and 35 to B. But she only had 20 so she couldn't do it. *[What if they had started the game even?]*

19. Amelia took tiles from Ramon's collection first, and then Keiko took half of the remaining tiles. Working backward we will start with the fact that Ramon had 11 left in the end. Just before this Keiko took half of the tiles, so Keiko took 11 from the 22 tiles that Ramon had at that time. Before this, Amelia took 13, so Ramon started with 22 + 13 = 35 tiles.

21. The student created the diagram by letting a single square represent the second number. Since the first number is twice the second number, the first number is represented by two squares. The third number is twice the first number, so it needs to be four squares. Each square represents the same number. There are a total of seven squares with a sum of 112, so each square represents 112 ÷ 7 = 16. The first number is 32, the second one is 16, and the third one is 64.

23. The student is using a drawing to represent the information. There are various ways the student may have used this diagram to help solve the problem. One way is to notice that if we replace a doughnut with a cup of coffee then the cost goes up by 10 cents. So a cup of coffee costs .10 more than a doughnut. Four doughnuts would cost .10 less than one cup of coffee and three doughnuts, so four doughnuts cost .80, and one doughnut costs 20 cents. This means that one cup of coffee costs 30 cents. These prices can be checked with the diagram of the original information.

25. *Solution 1: by making a drawing*

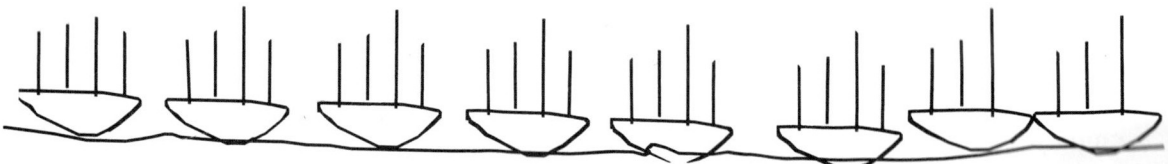

In this diagram, eight boats were drawn. First, three masts were placed on each boat, for a total of 24 masts. Then a fourth mast was added to boats until there were a total of 30 masts. So 6 boats had 4 masts and 2 boats had 3 masts.

*Solution 2: by making a table*

This table could be created by making sure that there are always a total of 8 boats. This is also how a guess and check strategy might work.

| boats with 4 masts | boats with 3 masts | total masts |
|---|---|---|
| 4 | 4 | 16 + 12 = 28 |
| 5 | 3 | 20 + 9 = 29 |
| 6 | 2 | 24 + 6 = 30 |

27. *Solution 1: by guessing and checking*
Suppose Claire got 10 free videos. That would mean that she paid for 30 videos. At $3 each this means she paid $90 for videos. But we were told that she paid $132, so the guess of 10 was too low. How much higher should we guess? 90 is about 2/3 of 132, so 15 might be a good guess. If Claire got 15 free videos that means she paid for 45. At $3 each that would cost her $135. This is very close, but she only spent $132. So, she paid for 44 videos. This is not quite enough to get 15 free, but it is enough for 14 free videos.

*Solution 2: by working backwards*
Since Claire paid $132 for her videos, she paid full price for 132 ÷ 3 = 44 videos. She gets a free video for every three that she pays for, so she has received 42 ÷ 3 = 14 free videos. If she pays for one more, she will get her 15$^{th}$ free video.

*[Note: A different interpretation could be made for this problem. Depending on the rules used by the video club, it may be that Claire needs to rent 3 videos all at once in order to receive the free one. If that were the case, we would not have enough information to solve the problem.]*

29. Solution 1: by making a drawing

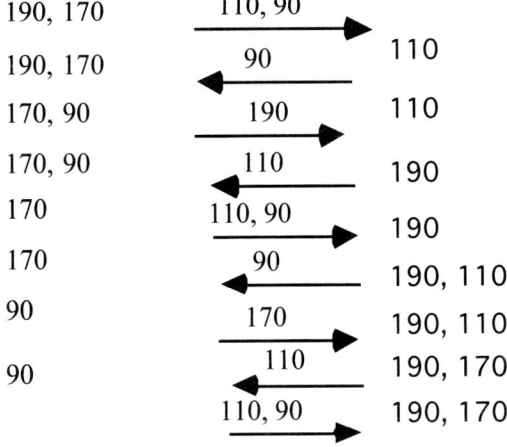

The drawing shows that the four people can cross the river in nine crossings. The 110 and 90 pound people are the only ones who can cross together, so they need to make the first trip and the last trip in order to accomplish the task in the minimum number of crossings.

*Solution 2: by using a model*
A similar method could be used to solve this problem by using four pieces of paper with the numbers 90, 110, 170, and 190 written on them. One could move them across a "river" following the rules given, experimenting until being convinced of a solution.

31. By a combination of working backward and guess and check:

    The numbers 6, 7, and 8 total to 21, so if one disk had a number that is one larger on the back we would get a total of 22. Suppose that the 6 has a 7 on the back. Then we could get 7, 7, and 8 for a total of 22. The smallest possible total needs to be 15. One way to get 15 is with a 6, 5, and 4. Let's try putting a 5 on the back of the 7 and a 4 on the back of the 8. Then our discs look like this:  **6 or 7;  7 or 5;  8 or 4.**

    Now check to see if we can get the totals we want.  $6 + 5 + 4 = 15$   $7 + 5 + 4 = 16$
    $6 + 7 + 4 = 17$   $7 + 7 + 4 = 18$   $6 + 5 + 8 = 19$   $7 + 5 + 8 = 20$   $6 + 7 + 8 = 21$
    $7 + 7 + 8 = 22$   It works!

    *[Note: The answer in the back of the text gives a different solution for this problem. This problem solver is not at all sure that these are the only two possible correct solutions. Can you find another one . . . or prove that these are the only ones?]*

33. Start both timers at the same time. The vegetables should be put in the steamer after the seven minute timer has just run out. At that point the 11 minute timer still has four minutes to go, so when it is finished the vegetables will have steamed for four minutes. Then turn over the 11 minute timer again and when it is finished the vegetables will have been steamed for 15 minutes.

35. These problems require careful reading and common sense thinking.

    a. If **you take** two apples, then **you** will have two apples. (Unless you already had some more.)

    b. All **but** nine died, so he still had nine left.

    c. "One is not a nickel" can be interpreted to mean that one of the coins is a nickel and one is not. So the coins are a quarter and a nickel.

    d. The cider costs 60 cents **more** than the bottle, so if the bottle was 10 cents, the cider would be 70 cents for a total of 80 cents. We need 6 cents more, so make the bottle 13 cents and the cider 73 cents.

    e. If it is a hole, it has no dirt in it. The amount of dirt removed is another question.

    f. If the hen weighs 3 pounds plus half its weight, we can think of the hen's weight as being divided into two equal parts. One half is 3 pounds and the other half is half it's weight. But the two parts have equal weight, so the other half must also be 3 pounds. The hen weighs 6 pounds.

    g. As long as there were no **men** playing baseball not a man would cross the plate. The players could have been women or children.

    h. Neither phrase is correct as long as the yolk didn't break. If uncooked the whites are clear and if cooked the whites are white.

## Section 1.2

1. a. This pattern has a core of three repeating characters. ▲+○

   b. This pattern repeats the core of the first four characters. ✖✖+✖

   c. One interpretation of this pattern is that it has a core which grows as follows: the first core is circle, star, 2 circles; the second core is circle, star, 3 circles; the third core is circle, star, four circles, etc.  ○✲○○○○○…

3. a. This is an arithmetic sequence with common difference three. So the next three numbers are 26, 29, and 32.

   b. The method of finite differences is helpful in this pattern. The sequence of differences between the numbers is 3, 3, 4, 4, 5, 5, 6. So to continue this pattern we want the next three differences to be 6, 7, and 7. That means that the next three numbers are 49, 56, and 63.

   c. Again looking at the differences between consecutive numbers in the sequence, we see 5, −2, 5, −2, 5, −2, 5 as the sequence of differences. Continuing this pattern gives 29, 34, and 32 as the next three numbers in the sequence.

5. The sequence of numbers of cannon balls in the first four figures shown is 1, 5, 14, 30. Using the method of finite differences gives us the sequence 4, 9, 16. These numbers may or may not look significant. If not, take finite differences again and get 5, 7. This means that the next number in the sequence 4, 9, 16 should be 16 + 9 = 25. The numbers 4, 9, 16, and 25 are the squares of the numbers 2, 3, 4, and 5. This makes sense with the cannon ball pattern because each new figure adds a larger base to the previous figure. Each base is a square layer of cannon balls. In the third figure the base is a 3 by 3 layer of 9 cannon balls. So the third figure contains 9 + 4 + 1 = 14 cannon balls. The sixth figure contains 36 + 25 + 16 + 9 + 4 + 1 = 91 cannon balls. The tenth pyramid consists of ten layers of cannon balls. The bottom layer contains 10 × 10 or 100 cannon balls. The next layer has 9 × 9, etc. until the top layer is a single cannon ball. So the tenth pyramid contains 100 + 81 + 64 + 49 + 36 + 25 + 16 + 9 + 4 + 1 = 385 cannon balls.

7. a. The sequence is arithmetic with a common difference of 3. Three additional cubes are added to each figure.

   b. Here is one description: the 20th figure has 39 cubes on the bottom and 19 cubes stacked on top of the first cube on the left. Another description: the 20th figure has 20 cubes stacked up and 38 cubes lined up next to the stack. Either way, there are 58 cubes. (Other descriptions are also possible.)

9. Multiplying the middle number by three will always give the sum of three consecutive numbers. If we call the middle number n, then the one to its left will be n − 1 and the one to its right will be n + 1. Adding the three numbers together gives n − 1 + n + n + 1. This sum is equal to n + n + n, which is the same as the product of 3 times n. 18 times 3 is 54, so 17 + 18 + 19 = 54.

11. The sum of the 9 numbers in a 3 by 3 array on a calendar will always be equal to 9 times the middle number. This can be seen using algebraic expressions as in #9, or we can look at it another way. In the array shown in the text with 11 in the middle there are four other pairs of numbers, each with sum 22 (or average 11). These four pairs are 3 and 19, 5 and 17, 4 and 18, 10 and 12. Note that these pairs consist of opposite corners, top and bottom, and the two sides. Why does it work this way? For example, the upper left corner of the array will always have a number that is 8 less than the middle number and the bottom right corner will always be 8 more than the middle. Similar relationships hold for the other pairs of numbers. To find a 3 by 3 array whose sum is 198 we would need the middle number to be $198 \div 9 = 22$. *[Could we always find such an array on any month's calendar?]*

13. $1^2 + 1^2 + 2^2 + 3^2 + 5^2 = 1 + 1 + 4 + 9 + 25 = 40 = 5 \times 8$
    $1^2 + 1^2 + 2^2 + 3^2 + 5^2 + 8^2 = 1 + 1 + 4 + 9 + 25 + 64 = 104 = 8 \times 13$
    $1^2 + 1^2 + 2^2 + 3^2 + 5^2 + 8^2 + 13^2 = 1 + 1 + 4 + 9 + 25 + 64 + 169 = 173 = 13 \times 21$

    This pattern shows that the sum of the squares of the first n Fibonacci numbers is equal to the product of the nth Fibonacci number times the next (i.e. the (n + 1)st) Fibonacci number. For example the sum of the squares of the first seven Fibonacci numbers is equal to the product of the seventh Fibonacci number times the eighth Fibonacci number. *[It might be an interesting exploration to try to figure out why this works.]*

15. If these are to be Fibonacci-type sequences, then any number in the sequence is found by adding the previous two numbers. So, in part a the two blanks before 11 might be 4 and 7, 3 and 8, 5 and 6, or any two numbers adding to 11. But the only pair of numbers that will work here and that also go with the 1 at the beginning of the sequence are 5 and 6. In part b, the blank between 14 and 20 has to be a 6, because the first two numbers must add to 20. This also confirms that the fourth number should be 26.

    Here are the sequences for parts a and b.
    a. 1, 5, 6, 11, 17, 28, 45, 73, 118, 191    b. 14, 6, 20, 26, 46, 72, 118, 190, 308, 498
    c. Exactly ten numbers of the sequence are given here, so we need to add them together.
       $14 + 6 + 20 + 26 + 46 + 72 + 118 + 190 + 308 + 498 = 1298$. The seventh number is 118.
       $11 \times 118 = 1298$, so it is true that the sum equals 11 times the seventh number.

    d. The first 10 terms of the Fibonacci sequence are 1, 1, 2, 3, 5, 8, 13, 21, 34, 55. Their sum is 143, which is equal to $11 \times 13$.

    e. One reasonable conjecture would be that the sum of the first 10 numbers of any Fibonacci-type sequence is equal to 11 times the seventh number in that sequence. *[What would you do if you wanted to prove or disprove your conjecture?]*

## 1.2 Patterns and Problem Solving

17. a. Here are the three given equations:

$$1 + 2 = 3$$
$$4 + 5 + 6 = 7 + 8$$
$$9 + 10 + 11 + 12 = 13 + 14 + 15$$

To continue this pattern, we write $\quad 16 + 17 + 18 + 19 + 20 = 21 + 22 + 23 + 24$
as the next equation so that it continues counting off the natural numbers and also adds one term in length to each side of the equation.

*[Do you notice anything about the first number on the left of each equation?]*

b. The first number in the first row is 1, which is the square of 1. The first number in the second row is 4, which is the square of 2. The third equation starts with 9, which is the square of 3, etc. So the 20$^{th}$ equation will start the sum with 20 squared, which is $20 \times 20 = 400$. Notice also how the number of terms grows on each side of each equation. The 1$^{st}$ equation has 2 terms equal to 1 term, the 2$^{nd}$ one has 3 terms equal to 2, the 3$^{rd}$ has 4 terms on the left and 3 on the right, etc. The 20$^{th}$ equation will have 21 terms on the left and 20 on the right. (We also could have found its length another way. The 21$^{st}$ equation needs to start with $21 \times 21 = 441$.) In any case, the 20$^{th}$ equation is:
$400 + 401 + 402 + \ldots + 420 + 421 = 422 + 423 + \ldots + 439 + 440$.

*[Why do you think this pattern works? You might try taking the first number on the left and distributing it among the other numbers on the left.]*

19. a. The third diagonal in Pascal's triangle contains the Triangular numbers, so the numbers in this diagonal are 1, 3, 6, 10, 15, 21, 28, 36, 45, 55, 66, 78, etc. where the difference between consecutive numbers keeps increasing by one. The tenth number in this sequence is 55. *[It is an interesting coincidence that the 10$^{th}$ Fibonacci number is also 55.]*

b. The first few numbers in the fourth diagonal are 1, 4, 10, 20, 35. Look at the differences between these numbers and also look back at part a. What do you notice? The *differences* between consecutive numbers in the fourth diagonal are consecutive triangular numbers. So the sequence continues with $35 + 21 = 56$ and then $56 + 28 = 84$, etc. The first ten numbers in the fourth diagonal are 1, 4, 10, 20, 35, 56, 84, 120, 165, 220.

21. The sum of each row of Pascal's triangle is a power of 2. The sum of the numbers in the 12$^{th}$ row is 2 to the 12$^{th}$ power, which is 4096.

23. a. This is an arithmetic sequence because there is a common difference. To obtain any number in the sequence, add 5 to the previous number. The sequence is 4, 9, 14, 19, 24, 29, 34, 39, 44, 49, 54, 59, etc. so the 12$^{th}$ number is 59. *[Could this have been found without writing out all the terms?]*

b. This is a geometric sequence because there is a common ratio. To obtain the next number, multiply the number by 2. The sequence is 15, 30, 60, 120, 240, 480, 960, 1920, 3840, 7680, 15360, 30720 so the 12$^{th}$ number is 30720.

c. This is an arithmetic sequence because it has a common difference. The numbers are decreasing, so the difference is negative. (Or think of subtracting instead of adding.) To obtain any number in the sequence, subtract 4 from (or add –4 to) the previous number. The sequence is 24, 20, 16, 12, 8, 4, 0, –4, –8, –12, –16, –20 so the 12$^{th}$ number is –20.

d. This is a geometric sequence because there is a common ratio. To obtain the next number, multiply the number by 3. The sequence is 4, 12, 36, 108, 324, 972, 2916, 8748, 26244, 78732, 236196, 708588 so the 12$^{th}$ number is 708588.

25. a. We can use the method of finite differences to find the next number in a diagonal of Pascal's triangle. For the first and second diagonals finite differences are not even needed (although they could be used). The first diagonal is all 1's and the second diagonal is the natural numbers (with constant finite difference 1). The third diagonal is the triangular numbers: 1, 3, 6, 10, 15, etc. with finite differences 2, 3, 4, 5, etc. so the next number after 15 is 15 + 6 = 21. The fourth diagonal starts with 1, 4, 10, 20, 35. Finite differences here are 3, 6, 10, 15. Notice that these are the numbers in the third diagonal. In the fourth diagonal the next number after 35 is 35 + 21 = 56. *[Can you see why the pattern works based on how Pascal's triangle is constructed?]*

b. What happens when we use finite differences on the sequence 1, 2, 4, 8, . . .?

1    2    4    8    16    32    64
   1    2    4    8    16    32
      1    2    4    8    16

No matter how many times we take differences we will always just have the same sequence again. So the method of finite differences is not helpful here.

27. a. Taking finite differences 3 times gives us the following:

1    2    7    22    53    106
   1    5    15    31    53
      4    10    16    22
         6    6    6

When we get down to this row where the difference is constant at 6, then we know that the next number in the second row up is 22 + 6 = 28. This means that the number after 53 is 53 + 28 = 81. Then this gives us the next number in the sequence. It is 106 + 81 = 187.

b. This sequence only requires two rows of finite differences before reaching a constant difference.

1    3    11    25    45    71
   2    8    14    20    26
      6    6    6    6

So the next number in the middle row is 32, and the next number in the sequence is 71 + 32 = 103.

29. A square number can be found by multiplying a natural number by itself. The sequence 1, 4, 9, 16, . . . can also be written as 1 × 1, 2 × 2, 3 × 3, 4 × 4, . . . So the next three numbers are 5 × 5, 6 × 6, and 7 × 7; or, 25, 36, 49. The 100th number in the sequence is 100 × 100, which is 10,000. This is why taking the second power is often called squaring.

31. a. Looking at the dimensions of the arrays in the sequence of oblong numbers helps to give us a clue. The first one is 1 × 2, the second is 2 × 3, the third is 3 × 4, and the fourth is 4 × 5. The fifth oblong number will be 5 × 6, or 30.

    b. The 20th oblong number will be 20 × 21, or 420.

    *[Do you see a connection between the oblong numbers and the triangular numbers? You might want to look at the arrays of dots.]*

33. a. The pentagonal numbers are 1, 5, 12, 22, 35, 51, . . . The first sequence of finite differences starts with 4, 7, 10, 13, 16. This is an arithmetic sequence because it has a common difference of 3.

    b. Yes, we can use this method to continue the sequence of pentagonal numbers. The difference between the 5th and 6th numbers was 16, so the difference between the 6th and 7th will be 19, the difference between the 7th and 8th will be 22, etc. The next few pentagonal numbers after 51 will be 70, 92, 92 + 25 = 117, 117 + 28 = 145, etc.

35. The reasoning used by the researchers in this study was inductive reasoning because the conclusions were based on observing a pattern that appeared in a sample population. These results were then extrapolated to the general population.

37. a. Here is a sketch of the 9th even number. It consists of two rows of nine; or nine sets of two.
    The 9th even number is 18.

    b. The 45th even number is 45 sets of two (or two 45s), so it is 45 × 2 = 90.

39. If we try the procedure on 8, we double 8 to get 16, then place them side by side to get 168. The number 168 is divisible by 7 because 7 × 24 = 168. So 8 is not a counterexample. If we try 12, we double 12 to get 24, then write the number 2412. This number is not divisible by 7.   344 × 7 = 2408 and 345 × 7 = 2415. So the number 12 is a counterexample, which shows that this procedure does not always give a number that is divisible by 7. The procedure does work for all one digit numbers.

41. a. To find a counterexample we just need to find one example where the suggested property does not work. So we are looking for two numbers that we can multiply so that the product is not divisible by 2. 7 × 4 = 28 is not a counterexample. 7 × 5 = 35 is a counterexample. *[What kinds of numbers do we need to use here to find counterexamples?]*

   b. We want to find a whole number greater than five which can not be written as the sum of two consecutive numbers and also can not be written as the sum of three consecutive numbers. Every odd number is the sum of two consecutive numbers. Look at the diagrams in #38 if you are not convinced of this. So we need to look in the even numbers. Can any even numbers be the sum of two consecutive numbers? Why not? <u>So for our counterexample we need to find an even number which can not be written as the sum of three consecutive numbers.</u> If we add any three consecutive numbers the result is always a multiple of three. To see this, draw a staircase pattern showing the sum of three consecutive numbers, like 6 + 7 + 8. This staircase can always be leveled off to a rectangle with width 3 simply by moving one square. For example, 6 + 7 + 8 = 7 + 7 + 7. So, if we can find an even number (greater than five) that is not a multiple of three then it will be our counterexample. Eight and ten are the smallest counterexamples. Fourteen is the next one.

43. a. The statement "the sum of any four consecutive whole numbers is divisible by 4" is false. A counterexample can be found in the first four whole numbers. 1 + 2 + 3 + 4 = 10. Ten is not divisible by 4. If we change the word "sum" to "product", then the statement will be true. We could prove this by noticing that any four consecutive numbers will contain one number that is a multiple of four. A product containing a factor which is a multiple of 47 will also be divisible by 4. There are other possible ways to correct the statement.

   b. This statement is true. Since we are only looking at 14 numbers, we could check the statement with each of the numbers.

45. a. By the end of the 2$^{nd}$ month the three first month members will have each brought in two new members, so that there will be 9 members. During the 3$^{rd}$ month each of the 9 members brings in two more members. Counting each old member with their two friends as a group of three, we see that there are 9 × 3 = 27 members. In fact, the system results in tripling the membership each month. So, after 6 months there will be 3 × 3 × 3 × 3 × 3 × 3 = 729 members. This product can also be written using an exponent. $3^6$ = 729.

   b. After 1 year (12 months) there will be $3^{12}$ = 531,441 members.

47. a. On the 12$^{th}$ day there were 12 + 11 + 10 + 9 + 8 + 7 + 6 + 5 + 4 + 3 + 2 + 1 = 78 gifts given. This is the 12$^{th}$ triangular number. It can also be found by using a staircase as in Sec. 1.1.

   b. The total number of gifts received during all 12 days is the sum of the first 12 triangular numbers, which is 1 + 3 + 6 + 10 + 15 + 21 + 28 + 36 + 45 + 55 + 66 + 78 = 364.

49. This problem will be easier to solve if we first look at a simpler problem. Suppose that the refrigeration car was the 3$^{rd}$ from the front and the 2$^{nd}$ from the end. Then the train would look like this: xxRx. The R is 3$^{rd}$ from the front and 2$^{nd}$ from the back. This train has 4 cars, which is 2 + 1 + 1 (or, 3 + 2 – 1). If the refrigeration car is the 147$^{th}$ from the front, then there are 146 others in front of it. And if it is 198$^{th}$ from the back, then there are 197 others behind it. So there are 146 + 1 + 197 = 344 cars total. (Or, 147 + 198 – 1 = 344 cars).

51. First look at the cards numbered from 1 to 100. Between cards number 1 and number 59, there are 6 cards containing a 6 (6, 16, 26, 36, 46, and 56). Each of cards 60 through 69 contains at least one 6, so this is 10 more cards. Also cards 76, 86, and 96 each contain a 6. There are a total of 19 cards between 1 and 100 containing at least one 6. The counts will be the same for the 100's, 200's, 300's, and 400's. So there are a total of 19 × 5 = 95 cards with a 6.

## Section 1.3

1. a. The variable x represents the depth. By replacing x with 10 in the expression .43x + 14.7, we find that the pressure at a depth of 10 feet is .43(10) + 14.7 = 4.3 + 14.7 = 19 pounds per square inch. If x is 100 we get a pressure of 43(100) + 147 = 4.3 + 147 = 151.3 pounds per square inch. To find the pressure at the surface we substitute 0 for x and get 43(0) + 14.7 = 0 + 14.7 = 14.7 pounds per square inch.

   b. Here we replace x with the number of chirps. For 20 chirps per minute, the temperature is 20/4 + 40 = 5 + 40 = 45 degrees. For 100 chirps per minute, the temperature is 100/4 + 40 = 25 + 40 = 65 degrees.

3. a. To find the cost of all of the main floor seats we would multiply the number of main floor seats by $28. An expression for the cost is 28m.

   b. To find the cost of all of the main floor seats we would multiply the number of main floor seats by $19. An expression for the cost is 19b.

   c. If the total for the main floor seats was greater than the total for the balcony seats then subtracting the total cost for the balcony seats from the total cost for the main floor seats will give us the positive difference. Using the answers from parts a and b, we get the expression 28m – 19b.

5. a. The correct equation is s = 6p. If there are 6 students for each professor, multiplying the number of professors by 6 will give the number of students. For example, if there are 10 professors, then the equation s = 6p says that there are 60 students.

   b. If one were to translate the verbal statement "6 times as many students as professors" into symbols without thinking too hard about what the symbols say, it might be easy to translate 6 <u>times</u> as many students to the symbols 6s. The thinking might be "6 times the students equals the professors" but that is not what the original statement said.

7.  a.  There are four more chips on the right side than on the left side, so we need to have two chips in each box for the scale to balance. If x represents the number of chips in one box, then the balance scale can be represented by the equation $2x + 5 = 9$.

    b.  Removing one box from each side of the scale we have 2 boxes and 4 chips on the left and 11 chips on the right side. Now remove 4 chips from each side. The scale still balances and there are 2 boxes on the left balancing 7 chips on the right. Each box must contain 3.5 chips. The balance scale can be represented by the equation $3x + 4 = x + 11$.

9.  a.  Step 1: Simplification of the left side using the distributive property.
        Step 2: Addition Property of Equality. (30 was added to both sides)
        Step 3: Simplification on both sides.
        Step 4: Subtraction Property of Equality. (7x was subtracted from both sides)
        Step 5: Simplification on both sides.
        Step 6: Division Property of Equality. (Both sides divided by 5)
        Step 7: Simplification on both sides.

    b.  Step 1: Simplification of the left side.
        Step 2: Subtraction Property of Equality. (2 was subtracted from both sides)
        Step 3: Simplification on both sides.
        Step 4: Division Property of Equality. (Both sides divided by 33)
        Step 5: Simplification on both sides.

11. The solutions of these equations given below show only one possible track. There are other correct ways to do each of these. For example, in part a we could start by subtracting 17x from both sides first.

    a.  
    | | |
    |---|---|
    | $43x - 281 = 17x + 8117$ | |
    | $43x - 281 + 281 = 17x + 8117 + 281$ | Addition Property |
    | $43x = 17x + 8398$ | Simplification |
    | $43x - 17x = 17x + 8398 - 17x$ | Subtraction Property |
    | $26x = 8398$ | Simplification |
    | $26x/26 = 8398/26$ | Division Property |
    | $x = 323$ | Simplification |

    b.  
    | | |
    |---|---|
    | $17(3x - 4) = 25x + 218$ | |
    | $51x - 68 = 25x + 218$ | Simplification |
    | $51x - 68 + 68 = 25x + 218 + 68$ | Addition Property |
    | $51x = 25x + 286$ | Simplification |
    | $51x - 25x = 25x + 286 - 25x$ | Subtraction Property |
    | $26x = 286$ | Simplification |
    | $26x/26 = 286/26$ | Division Property |
    | $x = 11$ | Simplification |

c.  $56(x + 1) + 7x = 45{,}353$
$56x + 56 + 7x = 45{,}353$     Simplification (Distributive Property)
$63x + 56 = 45{,}353$     Simplification
$63x + 56 - 56 = 45{,}353 - 56$     Subtraction Property
$63x = 45{,}297$     Simplification
$63x/63 = 45{,}297/63$     Division Property
$x = 719$     Simplification

d. $3x + 5 = 2(2x - 7)$
$3x + 5 = 4x - 14$     Simplification
$3x + 5 - 3x = 4x - 14 - 3x$     Subtraction Property
$5 = x - 14$     Simplification
$5 + 14 = x - 14 + 14$     Addition Property
$19 = x$     Simplification

13. a. If we remove 5 chips from each side, then the right side will still have the same amount of excess weight. Then there are 2 boxes on the left and 7 chips on the right. The right side will stay heavier as long as there are less than 3.5 chips in each box.
An inequality to represent the original situation is $2x + 5 < 12$.
An inequality representing the solution is $x < 3.5$

b. If we remove 2 chips from each side, then the left side will still have the same amount of excess weight. Then there are 3 boxes on the left and 9 chips on the right. The left side will stay heavier as long as there are more than 3 chips in each box.
An inequality to represent the original situation is $3x + 2 > 11$.
An inequality representing the solution is $x > 3$.

15. a. Step 1: Subtraction Property of Inequality. ($2x$ was subtracted from both sides)
Step 2: Simplification on both sides.
Step 3: Subtraction Property of Inequality. (11 was subtracted from both sides)
Step 4: Simplification on both sides.
Step 5: Division Property of Inequality. (Both sides divided by 4)
Step 6: Simplification on both sides.

b. Step 1: Subtraction Property of Inequality.
Step 2: Simplification on both sides. (Note that $2x - 3x = -1x$, or $-x$)
Step 3: Multiplication Property of Inequality. (Both sides Multiplied by $-1$)
    *Note that the inequality sign changes direction here.*
Step 4: Simplification on both sides.

17. a.
$$6(x + 5) > 11x$$
$$6x + 30 > 11x$$  Simplify using distributive property
$$6x + 30 - 6x > 11x - 6x$$  Subtraction Property
$$30 > 5x$$  Simplification
$$30/5 > 5x/5$$  Division Property of Inequality
$$6 > x \quad (\text{or, } x < 6)$$  Simplification

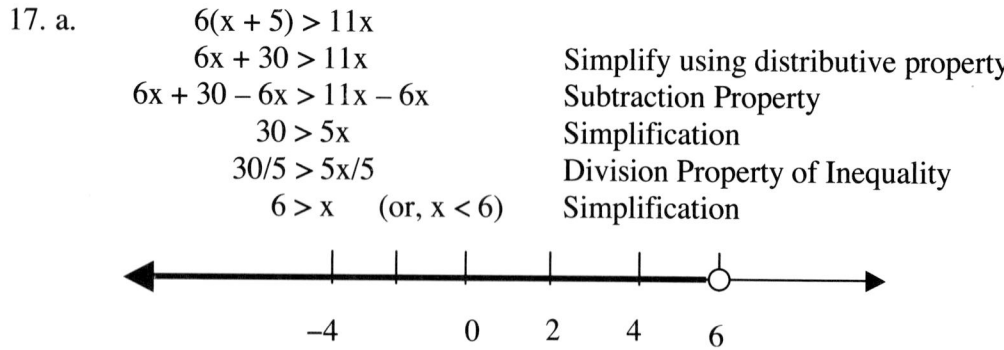

b.
$$5(x + 8) - 6 > 44$$
$$5x + 40 - 6 > 44$$  Simplification
$$5x + 34 > 44$$  Simplification
$$5x + 34 - 34 > 44 - 34$$  Subtraction Property
$$5x > 10$$  Simplification
$$5x/5 > 10/5$$  Division Property
$$x > 2$$  Simplification

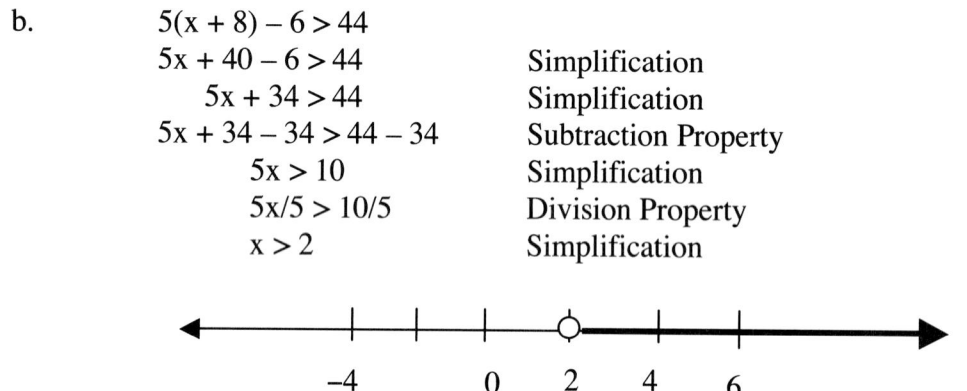

19. a. Each post card costs $.24. So if the number of post cards that Marci wrote is x, then the total cost to mail the post cards is .24x.

b. Marci sent a total of 18 post cards and letters. If we know the number of post cards, then we can find the number of letters by subtracting this amount from 18. For example, if she sent 10 post cards then she sent 18 – 10 = 8 letters. Since x is the number of post cards she sent, 18 – x is an algebraic expression for the number of letters she sent.

c. The cost for the letters is .39(18 – x) and the cost for the post cards is .24x so the total cost is .39(18 – x) + .24x.

d. Since the total cost is $5.22, we can write the equation .39(18 – x) + .24x = 5.22. Simplifying the left side by distributing the multiplication and combining like terms gives 7.02 – .15x = 5.22. Subtracting 7.02 from both sides gives –.15x = –1.80. Dividing both sides of this equation by ⁻.15 gives us x = 12.
So Marci mailed 12 postcards.

21. a. If x is the number of compact discs, then at $10.50 each their cost is 10.50x.

   b. There were three more tapes than discs, so there were x + 3 tapes.

   c. At $8 each, the total cost of the tapes was 8(x + 3). Note that it is important to put the x + 3 in parentheses here.

   d. The sum of the costs can be represented by 10.50x + 8(x + 3). We want this sum to be less than $120, so we want the solutions to the inequality 10.50x + 8(x+3) < 120.
   Simplify the left side to get  18.50x + 24 < 120.
   Subtract 24 on both sides  18.50x < 96.
   Divide both sides by 18.5  x < 5.189189 . . .
   To spend less than $120 Merle had to buy 5 or fewer compact discs.
   Check: If he bought 5 CD's, then he bought 8 tapes.
   This would cost 10.50(5) + 8(8) = $116.50.

23. Since we want to find the length of a side of the square region it makes sense to try letting x represent the length of a side of the square. Marcia put a fence around the four sides of the square, so she used 4x feet of fencing. She had 110 feet of fence left over, so she started with 4x + 110 feet of fencing. But we also know that she started with 350 feet of fence.
   So we know that  4x + 110 = 350.
   Solve by subtracting 100 from both sides to get  4x = 240
   And divide both sides by 4 to get  x = 60.
   The length of a side of the square region was 60 feet.
   *Note: Other lines of reasoning might lead to starting with the equations*
   *4x = 350 – 110  or  4x = 240.*

25. Another way to say the statement given is "The sum of a number and 14 is less than 3 times the number". Translating this phrase into symbols gives us  x + 14 < 3x.
   Subtract 1x from both sides to get  14 < 2x.
   Divide both sides by 2  7 < x.
   The solution says that the statement is true whenever x > 7.
   (Try this with some numbers.)

27. a. If n is the first number, then the next number is one more than n, or n + 1. The number following n + 1 is n + 2, and the next number is n + 3. So n, n + 1, n + 2, and n + 3 are four consecutive numbers.

   b. The sum of four consecutive numbers in which the first number is n could be written as n + (n + 1) + (n + 2) + (n + 3). The parentheses are optional in this expression.

   c. Since this sum is to have a value of 350, we can solve the equation
      n + n + 1 + n + 2 + n + 3 = 350.
   The four n's added together equal 4n and 1 + 2 + 3 = 6, so an equivalent equation is
      4n + 6 = 350.
   Subtracting 6 from each side we get, 4n = 344. Dividing by 4 on each side gives n = 86.

   d. If 350 could be written as the sum of three consecutive numbers then we could solve the equation n + n + 1 + n + 2 = 350 and obtain a whole number for n.
   But this gives 3n + 3 = 350, or 3n = 347. Since 347 is not evenly divisible by 3, there are no three consecutive numbers with sum 350. On the other hand, the equation
   n + n + 1 + n + 2 + n + 3 + n + 4 = 350 does have a whole number solution for n.
   Find it! So, there are 5 consecutive numbers that sum to 350. *[How about 6? or 7?]*

29. For example, start with the number 17. Adding 221 gives 238. Multiplying 238 by 2652 gives 631,176. Subtract 1326 to get 629,850. Divide by 663 to get 950. Subtract 870 to get 80. Dividing by 4 gives us 20. Then if we subtract our original number of 17 we have 3.
    Try starting with a different number and see if you also end up with 3.
   Why does this work? Suppose we call our original number x. Then after the first step we have x + 221. Next is 2652(x + 221); then 2652(x + 221) − 1326;

   then [ 2652(x + 221) − 1326] ÷ 663; then [( 2652(x + 221) − 1326) ÷ 663] − 870;

   then [( 2652(x + 221) − 1326) ÷ 663] − 870;

   then {[( 2652(x + 221) − 1326) ÷ 663] − 870} ÷ 4;

   Finally we subtract x to get ({[( 2652(x + 221) − 1326) ÷ 663] − 870} ÷ 4) − x.

   Simplifying this expression starting with the innermost parentheses gives us:

   $$({[( 2652x + 586092 - 1326) \div 663] - 870} \div 4) - x$$
   $$= ({[( 2652x + 584766) \div 663] - 870} \div 4) - x$$
   $$= ({(4x + 882) - 870} \div 4) - x$$
   $$= ((4x + 12) \div 4) - x$$
   $$= (4x \div 4) + (12 \div 4) - x$$
   $$= x + 3 - x$$
   $$= 3$$

   To really understand how this works be sure that you can justify each step to yourself. *[Try creating your own number trick.]*

31. The first scale shows that two cubes and a bolt balance with 8 nails. The second one shows that one cube balances a bolt and a nail. Since both scales are balanced, if we put the items from the left pan of the second scale onto the left pan of the first scale and put the items from the right pan of the second scale onto the right pan of the first scale, it will still balance.
This means that 3 cubes and a bolt balances with 9 nails and a bolt. Remove a bolt from each side to see that 3 cubes balances 9 nails. Then one cube will balance with 3 nails.

33. If x represents the distance from town C to town D, then 10x represents the distance from B to C, since it is 10 times farther. And, it is 10 times further yet from A to B, so the distance from A to B is 10(10x), or 100x. The total distance from town A to town D is the sum of these three distances. This sum is x + 10x + 100x. We also know that this total distance is 1332 miles. So we can write the equation x + 10x + 100x = 1332. To solve this equation we simplify the left side to be 111x and then divide both sides by 111 to get x = 12. This is the distance from C to D. The distance from town A to town B is 100x, which is 1200 miles.

35. There are various ways to see this pattern. You might see three sets of the figure number, with some overlap. For example, in the 4$^{th}$ figure there are three sets of 4 tiles, two sets going vertically and one set going horizontally. But there are two tiles that are counted twice using this method. So there are 3(4) – 2 tiles. The 10$^{th}$ figure would have 3(10) – 2 tiles and the nth figure would have 3n – 2 tiles. Or, you might look at the 4$^{th}$ figure and see three sets of 3 tiles, plus one extra tile. Then there are 3(3) + 1. The 10$^{th}$ figure then would have 3(9) + 1 tiles and the nth figure would have 3(n – 1) + 1 tiles. (Notice that this expression can be written as 3n – 3 + 1, which is also equal to 3n – 2.

So we will try to solve the equation 3n – 2 = 8230. Adding 2 to both sides gives 3n = 8232. Dividing both sides by 3 gives n = 2744. So the 2744$^{th}$ figure in the pattern would have 8230 tiles.

37. There are several different correct ways to view this pattern. For example, in describing the 3$^{rd}$ figure in the pattern, one person might see a 5 by 5 square of yellow tiles surrounded by a border of red tiles. Another person might see 7 reds on top, 7 on the bottom, and 5 additional on each side, with 5 rows of 5 yellow tiles in the middle. A third viewpoint might be a 7 by 7 square of red tiles with the 5 by 5 tiles in the center changed to yellow. Other correct views are possible.

   a. <u>Solution 1</u>: The fourth figure would have a 7 × 7 square of yellow tiles in the middle, surrounded by red tiles. There would by 4 × 8 = 32 red tiles. This method counts the top left corner with the top row, the top right corner with the right side, the bottom right corner with the bottom and the bottom left corner with the left side. Then the fifth figure would have 9 × 9 = 81 yellow tiles, surrounded by 4 × 10 = 40 red tiles.

   <u>Solution 2</u>: The fourth figure would have 9 reds on top, 9 on the bottom, and 7 on the left and 7 on the right, for a total of 9 × 2 + 7 × 2 = 32 reds tiles. It has 7 × 7 = 49 yellow tiles. The fifth figure has 11 × 2 + 9 × 2 = 40 red tiles and 9 × 9 = 81 yellow tiles.

   <u>Solution 3</u>: The fourth figure has 9 × 9 – 7 × 7 = 81 – 49 = 32 red tiles. It has 7 × 7 = 49 yellow tiles. The fifth figure has 11 × 11 – 9 × 9 = 121 – 81 = 40 red tiles and 9 × 9 = 81 yellow tiles.

   The first figure has 8 red tiles and 1 yellow tile. The second has 16 red and 9 yellow. The third has 24 red and 25 yellow.

   b. Here are three descriptions of the 100$^{th}$ figure, using the thinking of the 3 solutions in a:

   **1.)** Figure 100 has a 199 × 199 square of yellow tiles in the middle. (199 is the 100$^{th}$ odd number.) It is surrounded by a border of 4 × 200 red tiles. There are 800 red tiles and 39,601 yellow tiles. The total number of tiles is 800 + 39,601 = 40,401.
   **2.)** Figure 100 has 201 reds on top, 201 on the bottom, and 199 on each side. There are 2 × 201 + 2 × 199 = 800 red tiles. It has 199 × 199 = 39,601 yellow tiles. The total number of tiles is 800 + 39,601 = 40,401.
   **3.)** Figure 100 has 201 × 201 – 199 × 199 = 40,401 – 39,601 = 800 red tiles. It has 199 × 199 = 39,601 yellow tiles. The total number of tiles is 201 × 201 = 40,401.

   c. <u>Solution 1</u>: The n$^{th}$ figure has $(2n – 1)(2n – 1)$ or $(2n – 1)^2$ yellow tiles.
   It has 4 × 2n or 8n red tiles. There are a total of $(2n – 1)^2 + 8n$ tiles.

   <u>Solution 2</u>: The n$^{th}$ figure has $(2n – 1)(2n – 1)$ or $(2n – 1)^2$ yellow tiles.
   It has 2n + 1 reds on top, 2n + 1 on the bottom, and 2n – 1 on each side for a total of 8n reds. There are a total of $(2n – 1)^2 + 8n$ tiles.

   <u>Solution 3</u>: The nth figure has $(2n – 1)(2n – 1)$ or $(2n – 1)^2$ yellow tiles.
   It has $(2n + 1)^2 – (2n – 1)^2$ red tiles.
   There are a total of $(2n – 1)^2 + (2n + 1)^2 – (2n – 1)^2 = (2n + 1)^2$ tiles.

   *[Note: If these algebraic expressions in part c do not make sense to you, try thinking of n as being 100 and compare the part c answers with those in part b.]*

## Chapter 1 Test

1. Polya's four steps for problem solving are: Understanding the problem, Devising a plan, Carrying out the plan, and Looking back. These are discussed in Section 1.1 of the text.

2. The eight problem solving strategies introduced in this chapter are:

    Making a drawing    Using a model
    Guessing and checking    Working backward
    Making a table    Finding a pattern
    Using a variable    Solving a simpler problem

3. $1 + 2 = 3$
   $1 + 2 + 5 = 8$
   $1 + 2 + 5 + 13 = 21$
   $1 + 2 + 5 + 13 + 34 = 55$
   Each of these sums is itself a Fibonacci number. Each one is the next Fibonacci number after the last number being added. Another observation is that each answer in the pattern is the sum of the previous answer and the last number being added (e.g., $55 = 21 + 34$).

4. Here is row nine of Pascal's Triangle: 1 9 36 84 126 126 84 36 9 1. The sum of these numbers is 512. Another way to find the sum of the ninth row of Pascal's triangle is to recall that each row of Pascal's triangle sums to a power of 2.
   The sum of row nine is $2^9 = 512$.

5. a. This is a geometric sequence with a common ratio of 3. The next term is $81 \times 3 = 243$.

    b. This is an arithmetic sequence with a common difference of 3. The next term is 18.

    c. This is an arithmetic sequence with a common difference of 6. The next term is 30.

    d. This is a sequence of consecutive square numbers. The next square number is 36. The pattern for this sequence can also be found by taking finite differences.

    e. Taking finite differences we get:    3    5    11    21    35
                                               2    6    10    14
                                                  4    4    4
    So the next number is $35 + 18 = 53$.

6. In problem 5, sequence a is geometric and sequences b and c are arithmetic. Sequences d and e are neither arithmetic nor geometric because in both of them there is neither a common difference nor a common ratio.

22        *Chapter 1  Problem Solving*

7.  a.  Taking finite differences we get:    1     5     14     30     55
                                                 4      9     16     25
                                                    5     7     9
                                                       2     2

    Continuing the pattern to the right, the third row will be 5, 7, 9, 11, 13, 15; and the second row will be 4, 9, 16, 25, 36, 49, 64. To get the next numbers after 1, 5, 14, 30, 55, we can add using the row two numbers: 55 + 36 = 91, 91 + 49 = 140, 140 + 64 = 204.

    b.  Taking finite differences we get:    2     9     20     35
                                                 7     11     15
                                                    4     4

    Continuing the pattern to the right, the second row will be 7, 11, 15, 19, 23, 27

    and the sequence will be 2, 9, 20, 35, 54, 77, 104.

8.  a.  The first four triangular numbers are shown in the figure below. The difference between successive triangular numbers increases by one with each new number in the sequence. The sequence of triangular numbers is 1, 3, 6, 10, 15, 21, 28, . . .
    The fifth triangular number is 15.

    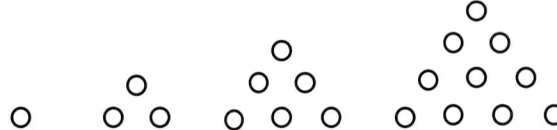

    b.  The square numbers can be drawn similarly, with the figures being squares instead of triangles. The sequence of square numbers is 1, 4, 9, 16, 25, 36, 49, . . .
    The fifth square number is 25.

    c.  The sequence of pentagonal numbers is 1, 5, 12, 22, 35, 51, 70, . . .
    The fifth pentagonal number is 35.
    *[Do you see a pattern in these three questions?  Can it be generalized?]*

9.  If we add the first seven consecutive whole numbers, 1 + 2 + 3 + 4 + 5 + 6 + 7 we get 28. This number is divisible by 4, so it is not a counterexample. But, 2 + 3 + 4 + 5 + 6 + 7 + 8 gives us 35, which is not evenly divisible by 4. So these seven consecutive whole numbers provide a counterexample. There are many other possible counterexamples too.

10. a.  Remove two boxes from each side and remove five chips from each side. The beam will remain balanced and we will see that one box balances with seven chips.

    b.  Remove two boxes from each side and remove seven chips from each side. The beam will remain in the same position and we will see that there are six chips on the left side and three boxes on the right. The three boxes will remain heavier than the six chips if each box contains more than two chips.

11. a.
$$3(x - 40) = x + 16$$
| | |
|---|---|
| $3x - 120 = x + 16$ | Simplification |
| $3x - 120 - x = x + 16 - x$ | Subtraction Property |
| $2x - 120 = 16$ | Simplification |
| $2x - 120 + 120 = 16 + 120$ | Addition Property |
| $2x = 136$ | Simplification |
| $2x/2 = 136/2$ | Division Property |
| $x = 68$ | Simplification |

b.
$$4x + 18 = 2(441 - 34x)$$
| | |
|---|---|
| $4x + 18 = 882 - 68x$ | Simplification |
| $4x + 18 + 68x = 882 - 68x + 68x$ | Addition |
| $72x + 18 = 882$ | Simplification |
| $72x + 18 - 18 = 882 - 18$ | Subtraction |
| $72x = 864$ | Simplification |
| $72x/72 = 864/72$ | Division |
| $x = 12$ | Simplification |

12. a.
$$7x - 3 < 52 + 2x$$
| | |
|---|---|
| $7x - 3 - 2x < 52 + 2x - 2x$ | Subtraction |
| $5x - 3 < 52$ | Simplification |
| $5x - 3 + 3 < 52 + 3$ | Addition |
| $5x < 55$ | Simplification |
| $5x/5 < 55/5$ | Division |
| $x < 11$ | Simplification |

b.
$$6x - 46 > 79 - 4x$$
| | |
|---|---|
| $6x - 46 + 4x > 79 - 4x + 4x$ | Addition |
| $10x - 46 > 79$ | Simplification |
| $10x - 46 + 46 > 79 + 46$ | Addition |
| $10x > 125$ | Simplification |
| $10x/10 > 125/10$ | Division |
| $x > 12.5$ | Simplification |

13. a. Step 1: Subtraction Property of Inequality. (2x was subtracted from both sides)
    Step 2: Simplification on both sides.
    Step 3: Addition Property of Inequality. (217 was added on both sides)
    Step 4: Simplification on both sides.
    Step 5: Division Property of Inequality. (Both sides divided by 13)
    Step 6: Simplification on both sides.

b. Step 1: Simplification using the distributive property.
   Step 2: Subtraction Property of Inequality. (15x was subtracted from both sides)
   Step 3: Simplification on both sides.
   Step 4: Division Property of Inequality. (Both sides Divided by 23)
   Step 5: Simplification on both sides.

14. Suppose that the fence was only 40 feet long.  O—O—O—O—O
    This diagram shows that a 40 foot long straight fence that begins and ends with a post and has posts that are 10 feet on center requires 5 posts. A 30 foot fence would need 4 posts. A 100 foot fence would need 11 posts. A 2000 foot fence would need 201 posts, one every 10 feet and one extra so that both ends have a post. This solution used the strategies of looking at a simpler problem and making a drawing. Other strategies could also be used.

15. Right before finishing with 170 chips, Pauli won 80 chips, so before winning these 80 chips she had 170 – 80 = 90 chips. The step that took her to 90 was losing half of her total chips, so she must have had double 90, or 180 chips before losing half of them. To get to 180, she won 50 chips, so before winning those 50, she had 180 – 50 = 130. In the first round she lost half her chips, which took her to the 130 position. So at the beginning of the game she had 130 × 2 = 260 chips. This solution uses the strategy of working backward. That seems to be the most natural approach for this problem, but other strategies could be used, such as a combination of guessing and checking and using a table.

16. One approach to solving this problem is to notice that the diagram of the tower is a representation of triangular numbers. The number of tiles required to build a tower with 25 tiles along the base and 25 rows of tiles will be the $25^{th}$ triangular number. The sequence of the first 25 triangular numbers is: 1, 3, 6, 10, 15, 21, 28, 36, 45, 55, 66, 78, 91, 105, 120, 136, 153, 171, 190, 210, 231, 253, 276, 300, 325. So it would take 325 tiles. If we have enough experience with the triangular numbers, we might recall that the $25^{th}$ triangular number is equal to (25 × 26) ÷ 2 which is 325. The $n^{th}$ triangular number is n(n + 1) ÷ 2. The strategies used here included finding a pattern and solving a simpler problem. Other useful strategies might include using a model, making a drawing, or making a table.

17. a. This problem seems to be a good one in which to use the finding a pattern strategy, since we are shown a pattern. One might look for a visual pattern or a pattern in the numbers of dots. The numbers are consecutive multiples of 4, so the fourth square should have 4 × 4 = 16 dots. There are a number of different ways to view the visual pattern. Counting the dots on the top, bottom, and sides of each square, one might count the dots in the fourth square as 5 on top, 5 on the bottom, and 3 between on each side. Or, one might count 3 on each of the four sides, plus the 4 corners. Another way to look at the pattern is to see a square array of dots with an inner square array removed. In this case one might count the dots in the fourth square as $5^2 - 3^2$. In any case, the fourth square has 16 dots.

    b. The method for determining the number of dots in the $50^{th}$ square will depend on the way one views the pattern. Four ways were mentioned in part a. Performing the calculations using each of the four methods in the same order as they are described above gives:
    
    $\quad\quad$ 4 × 50 = 200 $\quad\quad\quad\quad\quad\quad\quad$ 4 × 49 + 4 = 200
    $\quad\quad$ 51 + 51 + 49 + 49 = 200 $\quad\quad\quad$ $51^2 - 49^2 = 2601 - 2401 = 200$

    c. Each of the following is a correct expression for the number of dots in the $n^{th}$ square:

    $\quad\quad$ 4n $\quad\quad\quad\quad\quad\quad\quad\quad\quad\quad$ 4(n – 1) + 4
    $\quad\quad$ 2(n + 1) + 2(n – 1) $\quad\quad\quad\quad$ $(n + 1)^2 - (n - 1)^2$

18. Let's look at a simpler problem first.  Suppose that there are only four people around the table.  Call the people A, B, C, and D.  A shakes hands with B and with D.  B shakes hands with A and with C.  C shakes hands with B and with D.  D shakes hands with C and with A.  This looks like 8 handshakes, but each handshake has been listed twice here, so there are actually four handshakes.  Each person shakes the hand of two other people, but if we count each handshake separately this way, each shake gets counted twice.  So as long as there are at least three people, there will be exactly as many handshakes as there are people.  For 78 people there will be 78 handshakes.  For n people there are n handshakes.  This solution used the strategies of solving a simpler problem and of finding a pattern.  Another solution would be to represent the people around the table as vertices of a polygon, and to represent the handshakes as the edges of the polygon.  This might be done with a drawing or with a model.

19. The drawing shows that it would take nine times across the river for the small canoe to get everyone to the other side.  A similar procedure could be done using a model.

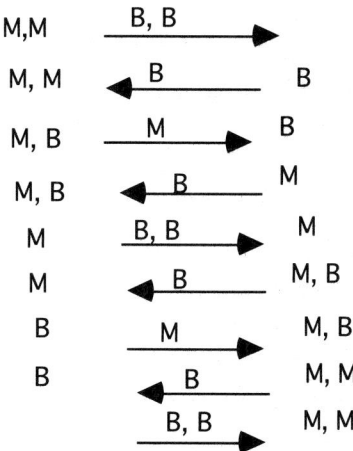

20. There are several ways to look at this pattern.  Here are two of them:
    I.   For example, in the 3$^{rd}$ figure notice that there are 3 + 2 tiles across the top and two sets of three tiles hanging below them for a total of 3 + 3 + (3 + 2) = 11 tiles.

    II.  Notice that the 3$^{rd}$ figure consists of a 4 by 5 rectangle with a 3 by 3 square removed.  So it contains 4(5) – 3(3) = 20 – 9 = 11 tiles.

    III. You may have another way to look at the pattern.

    a. <u>Using method I:</u>  The 5$^{th}$ figure has 5 + 5 + (5 + 2) = 17 tiles.
       The 150$^{th}$ figure has 150 + 150 + (150 + 2) = 452 tiles.
       <u>Using method II:</u>  The 5$^{th}$ figure has 6(7) – 5(5) = 42 – 25 = 17 tiles.
       The 150$^{th}$ figure has 151(152) – 150(150) = 22952 – 22500 = 452 tiles.

    b. <u>Using method I:</u>  The n$^{th}$ figure has n + n + (n + 2) = 3n + 2 tiles.
       <u>Using method II:</u>  The n$^{th}$ figure has (n + 1)(n + 2) – n(n) tiles.
       *[With some algebra techniques, we can see that these two expressions are equivalent.]*

# Chapter 2 Sets, Functions, and Reasoning
## Section 2.1

1. The system of notches on the bone could have been tally marks to count something. The marks do not necessarily mean that those who made the marks had names for the numbers.

3. Row 1 has marks arranged in the following groups: 3, 6, 4, 8, 10, 5, 5, 7.
   Row 2 has marks arranged in the following groups: 11, 13, 17, 19.
   Row 3 has marks arranged in the following groups: 11, 21, 19, 9.

   a. Row 1 could show a knowledge of multiplication by 2.

   b. The numbers in row 2 are the prime numbers between 10 and 20. The numbers in row three may have some connection with their proximity to 10 and 20. The numbers in row two add to 50. The numbers in row three add to 60.

5. Many different pairs of sets of these attribute pieces can be put into one-to-one correspondence. As long as two sets have the same number of pieces, they can be put into a one-to-one correspondence. Some examples of such pairs are: SY and LB; SY and LY; B and Y; H and R, etc.

7. There are many pairs of sets that are disjoint. Some examples are: Y and SB; SY and SB; SY and L; SB and L; R and T. Two sets are disjoint if they have no elements in common.

9. The pieces in the intersection Y ∩ L are LYT, LYR, and LYH. These are the pieces that have **both** of the attributes Yellow and Large.

11. a. This statement says that the large black hexagon is an element of the intersection of the set of large pieces and the set of yellow pieces. The LBH is large, but it is not yellow, so the statement is false. Since it is not in the yellow set, it is not in the intersection L ∩ Y.

    b. This statement says that the large yellow triangle is an element of the union of the set of hexagons and the set of large pieces. The LYT is large, so it is in the set of large pieces. This puts it in the union of H and L, so the statement is true. It does not need to be in H to be in the union.

    c. This statement says that the small yellow rectangle is in the complement of the set of hexagons. This statement is true because the SYR is not in the set of hexagons.

13. a. The SYH and SBH are the only pieces that are in the intersection of hexagonal **and** small.

    b. The pieces in not(hexagonal and small) are all of the pieces not listed in part a. These are:
       lbt, lbr, lbh, sbt, sbr, lyt, lyr, lyh, syt, syr.

    c. The pieces in the union small **or** yellow are all of the pieces that are either small or yellow or both. These pieces are: syt, syr, syh, sbt, sbr, sbh, lyt, lyr, lyh.

15. a. The elements in A ∩ C are 4 and 6. These are the only numbers that are listed in both of sets A and C.

   b. The complement of C consists of all members of the universal set that are not in C. These are 0, 1, 2, 7, 8. The set B includes some of these elements and also includes 3 and 5. So the elements of the union of the complement of C with B are 0, 1, 2, 3, 5, 7, 8.

17. a.

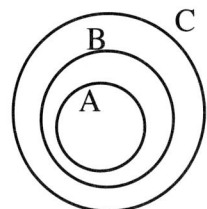

In this Venn diagram set C contains all of set B and set B in turn contains all of set A. So A is a subset of B and B is a subset of C.

   b.

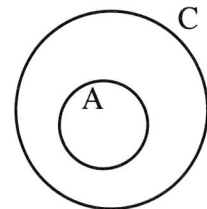 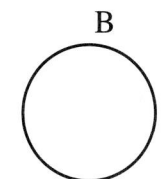

Now set A is a subset of set C, but the intersection of sets B and C is empty, so B and C do not intersect.

   c.

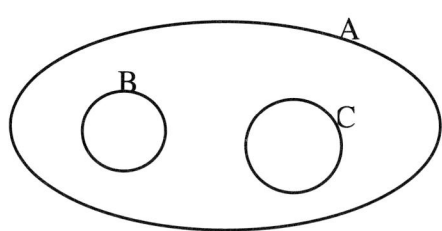

In this Venn diagram sets B and C do not intersect, but both are contained completely within set A, so the union is a subset of A.

19. a.  b.  c.

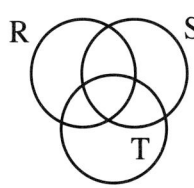 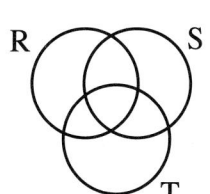 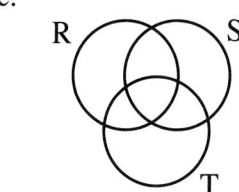

In part a we need to shade the intersection of T and R, so we shade only that parts of the circles that are contained in both sets simultaneously. In part b we are shading a union, so we include all of both sets T and R in the shaded area. For part c we first look at the union of sets R and S, then intersect that set with set T.

28  Chapter 2 Sets, Functions, and Reasoning

21. If set A has 15 elements and set B has 13 elements, then the union A ∪ B will have a maximum number of elements when the two sets are disjoint. In other words, when the two sets do not intersect there will be a total of 28 elements in the union. The intersection cannot contain more than 13 elements. A ∩ B will contain 13 elements when B is a subset of A. The diagrams below show both situations.

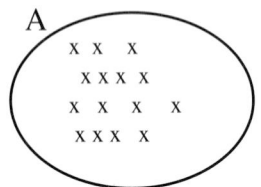 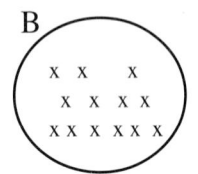   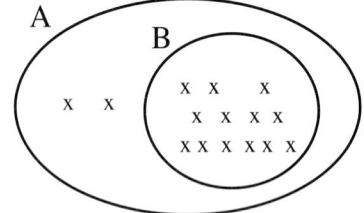

23. The set A ∩ B' means the intersection of set A with the complement of set B. So it includes the parts of set A that are not in set B.

25. The set A' ∪ B' contains all points that are not in A and also contains all points not in B. This is all points except those in A ∩ B.

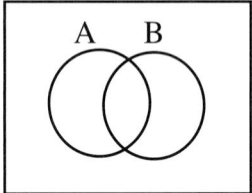

27. The shaded part in this problem is all of B except the part that is also in A. It is the intersection of B with the complement of A. In symbols this is A' ∩ B.

29. a. The intersection of sets S and B includes the regions d and c.

    b. The intersection of sets A and R includes the regions i and j.

31. If 4 people didn't play either instrument, then 11 people played at least one instrument. Since 7 played piano and 6 played guitar, there were 2 people who played both piano and guitar. The Venn diagram below helps to organize and keep track of the information.

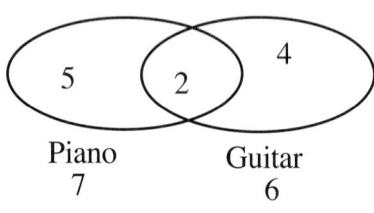 

33.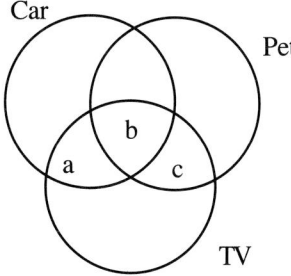

In the Venn diagram above we are looking for the number of people in the region labelled c. These are the people who have a TV and a pet, but not a car. The total number of people in regions b and c together is 1500. These are all those who have both a pet and a TV. There are 1100 people in region b, who have all three, a car, a pet, and a TV.
So there must be 1500 − 1100 = 400 people in region c.

35.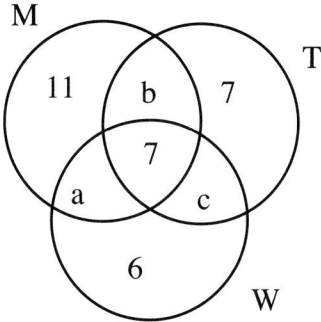

The Venn diagram above shows the information that is given in the problem. We also know that the total in circle M is 25, the total in T is 20, and the total in W is 16. If 12 students watched on both Monday and Tuesday, then region b must have 5 students. This means that region a has 25 − 11 − 5 − 7 = 2 students. Once we know the number in region a we can also find the number in region c. Region c has 16 − 6 − 2 − 7 = 1 student. All of the students watched TV on at least one of the days, so the total number of students is the sum of all the numbers in the regions of the diagram. This is 11 + 5 + 7 + 2 + 7 + 1 + 6 = 39 students. Notice that it is not the same as adding 25 + 20 + 16. Be sure you see why this is true.

37. In the diagram below, the numbers shown can be found in the given information. The intersection of DN and SG contains 34, and the total in DN is 129, so there are 95 people who read the Daily News only. The only missing quantity is the amount in region a. Since the poll was of 150 people, there must be 150 − 95 − 34 − 12 = 9 people in region a.
So there are a total of 34 + 9 = 43 Sun Gazette readers.

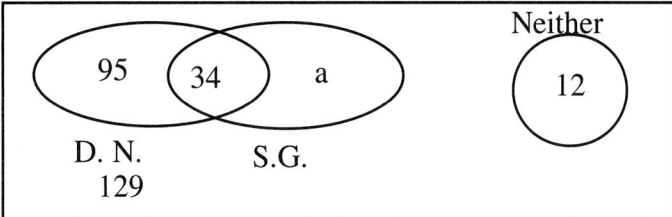

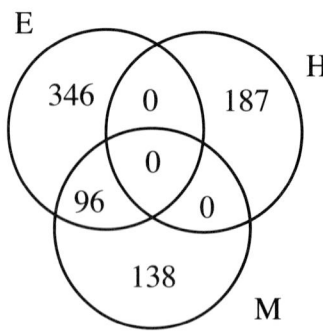

39. <u>Solution 1:</u> The diagram above shows the situation that would give the maximum total number of students while still satisfying the given information that there are 96 registered for both Math and English. The number of total students will be greatest when as few students as possible are taking more than one kind of course. For example, if we put some of the 96 students registered for both English and Math in the region in the center, where they are also registered for History, then the total will go down because the 187 taking only history will be reduced. The total number of students in the diagram above is 346 + 187 + 96 + 138 = 767.

<u>Solution 2:</u> There were a total of 442 + 187 + 234 = 863 class registrations. But we know that 96 people registered for both English and Math. If no one registered for both English and History, and no one registered for both History and Math, then the total number of different students is the maximum of 863 − 96 = 767, because then only 96 of the course registrations are of duplicate people. (One would hope that this would not be happening at a liberal arts college!)

41.

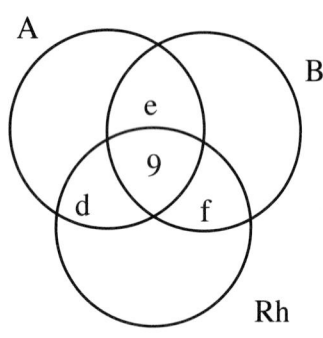

a. In this diagram we can fill in some more information from the chart. There were a total of 17 people with both antigens B and Rh, so region f contains 17 − 9 = 8 people with B+ blood type.

b. There were a total of 46 people with both antigens A and Rh, so region d contains 46 − 9 = 37 people with A+ blood type.

# Section 2.2

1. Move along the line from left to right. The numbers on the difficulty scale (the horizontal axis) are increasing. At the same time, the line is going downhill, so the motivation level is decreasing. As difficulty increases, motivation decreases. As difficulty decreases, motivation increases.

## 2.2 Functions and Graphs

3. If you double an element in the domain and then subtract one, you get the number in the range to which this element is assigned.

   a. Translating the sentence above into algebraic symbols we get $f(x) = 2x - 1$.

   b.
   | x | 1 | 2 | 3 | 4 | 5 | 6 | 7 | 8 | 9 | 10 |
   |---|---|---|---|---|---|---|---|---|---|----|
   | g(x) | 1 | 3 | 5 | 7 | 9 | 11 | 13 | 15 | 17 | 19 |

   c. The graph is plotted by setting up the horizontal axis to contain the numbers in the domain and the vertical axis for the range. A point is plotted at the intersection of a vertical line straight up from a number in the domain with a horizontal line straight over from the corresponding range coordinate.

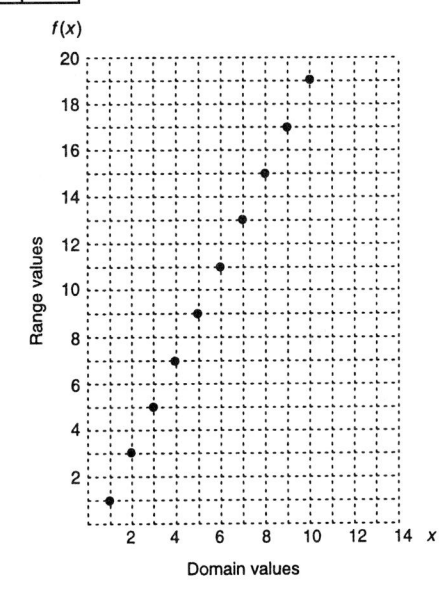

5. a. Yes, it is a function because everyone has exactly one birthday.

   b. Yes, it is a function because there is always exactly one number that is 10 times greater than any other number.

   c. No, this is not a function. A person may have more than one telephone number.

7. a. The rule is to add 17 to a number. Algebraically this is $f(x) = x + 17$.

   b. First we multiply by 3, then subtract 2. The rule written as a function is $g(x) = 3x - 2$.

9. a. This function takes as input the length of the side of a square. The output is then the area of the square. The range values will be the squares of the values in the domain. Since the domain is the numbers 1, 2, 3, 4, and 5, the corresponding range is 1, 4, 9, 16, and 25.

   b.

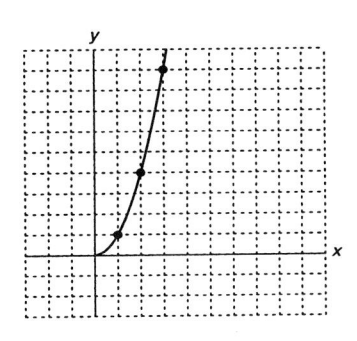

c. An equation for this function is $f(x) = x^2$

d. This function is nonlinear. We can tell it is nonlinear because its graph is not a line.

11. a. We will assume that the scale for each axis is one unit per space. Then we can count spaces to find the ratio of the change in y to the change in x. As y goes down two spaces, x goes to the right one space. The rise is –2 and the run is 1. So the slope is the ratio $\frac{-2}{1}$, or –2. The y-intercept is the point where the line crosses the y-axis, which is (0,2). Using the form y = mx + b, we can write the equation $y = -2x + 2$ to represent this line.

b. Here we count the rise to be 1 for a run of 4. Or, if we count between the two furthest points marked, we have a rise of 2 and a run of 8. Either way, the slope is the ratio $\frac{1}{4}$.
The y-intercept is the point (0,2), so an equation for the line is $y = \frac{1}{4}x + 2$.

13. a. The slope of line (i) is 10. We can see this from the equation, since the coefficient of x is 10 and if the graph were clear enough we could also see there that the ratio of the rise to the run is 10. For line (ii) the slope is 1.

b. As a line with positive slope becomes closer to vertical, its slope continues to increase. So we can draw a line a bit steeper than the line in (i) and have a line with greater slope.
For example, if we draw a line that rises 20 units for every run of one unit, we will have the line y = 20x.

c. There is no limit to how large the slope can be. Theoretically, we could make the rise as large as we want, and keep the run at one unit. As the line approaches a vertical line, its slope approaches infinity.

15. a. If each pound has a value of $1.50 then 5 pounds has a value of $1.50(5) = $7.50.

b. To convert dollars to pounds we need to divide by $1.50. For example, 3 dollars is worth 2 pounds, because 3 ÷ 1.5 = 2. To find the value of $22.50 we calculate 22.5 ÷ 1.5 = 17. So $22.50 has a value of 17 pounds.

c. The equation for this function is c(x) = 1.50x.
This could also be written c(x) = 1.5x, or c(x) = 3x/2.

d.

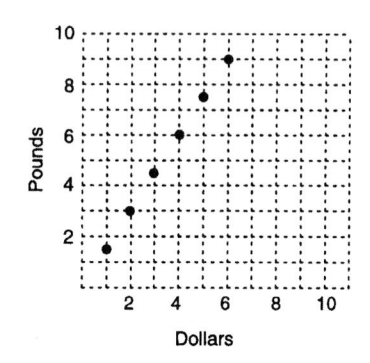

## 2.2 Functions and Graphs

17. a. State residents who are non members pay an initial fee of $2, so their cost for zero hours is $2. Their cost for 1 hour is $3. This corresponds to the middle line in the graph.

    b. For an 8 hour rental a club member pays $8 and an out-of-state resident pays 5 + 8 = $13. For 11 hours (or any number of hours) an out-of-state resident also pays $5 more than a member.

    c. On a sketch of the graphs, one can see that there is always a vertical distance of 5 between the top line and the bottom line no matter where on the graph this is measured.

    d. It will cost a state resident $2 more to rent a canoe for 9 hours than a club member.

19. a. The total cost is $40 plus $12 for each jack. If there are x jacks installed, then the cost is given by c(x) = 40 + 12x.

    b. The value of c(5) is found by replacing x with 5 in the equation.
       c(5) = 40 + 12(5) = 40 + 60 = $100. It costs $100 to have 5 jacks installed.

    c.

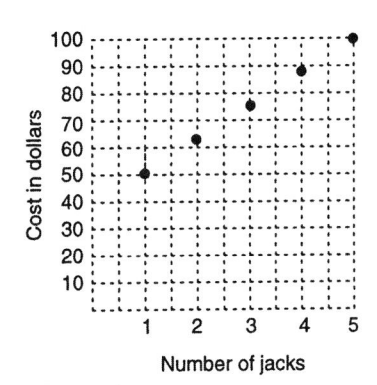

21. a. The total cost will be given by multiplying the number of hours used by 6 and then adding the initial charge of $15. An equation that does this is g(x) = 6x + 15. An equally good expression for the function would be g(x) = 15 + 6x.

    b. The value of g(14) is found by replacing x with 14 in the equation.
       g(14) = 6(14) + 15 = 84 + 15 = $99.

    c.

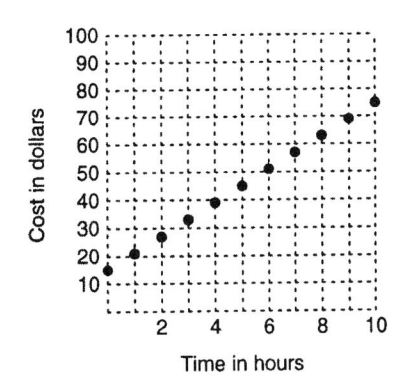

34                    *Chapter 2 Sets, Functions, and Reasoning*

23. a. From the 3rd to the 5th minutes the student is going at a constant speed of 12 miles per hour. This can be found from the graph by noting that when the time is between 3 and 5 the graph is a horizontal line at 12.

    b. The times when the student came to a stop are indicated by speeds of zero. The speed is zero at times 0, 10, and 20. The student was already stopped at the beginning, and came to a stop at 10 minutes and at 20 minutes.

    c. The speed was increasing during the times where the graph is going uphill. These are the intervals from 0 to 3 minutes, 6 to 7 minutes, 10 to 13 minutes, and 15 to 16 minutes.

25. a. Joel walked half the distance and then jogged the rest of the way. This would match graph IV, where the pace during the first part is steady and slow, and then is faster in the second part. Joan jogged all the way. This is graph II. It is at a constant rate, but not as fast as graph I. Graph I is Bob. Mary stopped for a time in the middle of her trip. This is indicated in graph III. (All of these answers assume scales on the graphs are all the same.)

    b. Joel took the greatest amount of time, because graph IV stretches the furthest to the right. Mary took almost as long.

    c. Mary lives the greatest distance from school because graph III goes the furthest on the distance scale.

    d. Bob took the least amount of time. Graph I is the shortest left to right.

    e. Joel lives closest to school. Graph IV does not reach as high vertically as the others.

27. a. In graph II the oven door was opened briefly during cooking. This would cause a quick drop in the temperature and then a resumption of the cooking temperature. This is seen in the V shape in the middle of the graph.

    b. Graph III shows the oven initially heated higher than needed, and then lowering slightly before leveling off.

    c. The last part of graph I shows a more or less constant temperature with slight variations.

    d. The first part of graph I shows that the oven was on at a low temperature at first.

29. This graph shows the speed of the roller coaster as a function of the position on the track. From A to B it is going slowly and then accelerates through C and D, hitting its top speed at the bottom of the hill at E and then slowing down again as it goes uphill to H.

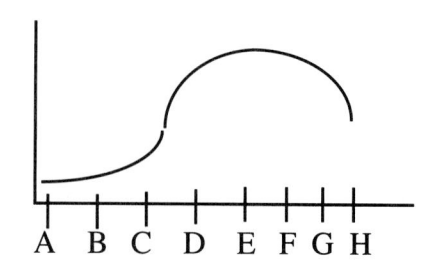

31. a. Since Pat gave Hal a 40 meter head start, Hal's graph starts at the point (0,40). Hal runs 3 meters per second, so after 10 seconds he has run 30 meters and is at the point (10,70). After 20 seconds he has run 60 meters and is at (20,100). Pat's graph starts at (0,0). Her speed is 4 meters per second, so after 10 seconds she has run 40 meters and is at the point (10,40). Her graph also goes through (15,60), (20,80), etc.

b. After 30 seconds, Pat is at (30,120) and Hal is at (30,130). Hal's distance of 130 meters comes from his headstart of 40 meters plus a distance of 90 meters run in 30 seconds at 3 meters per second.

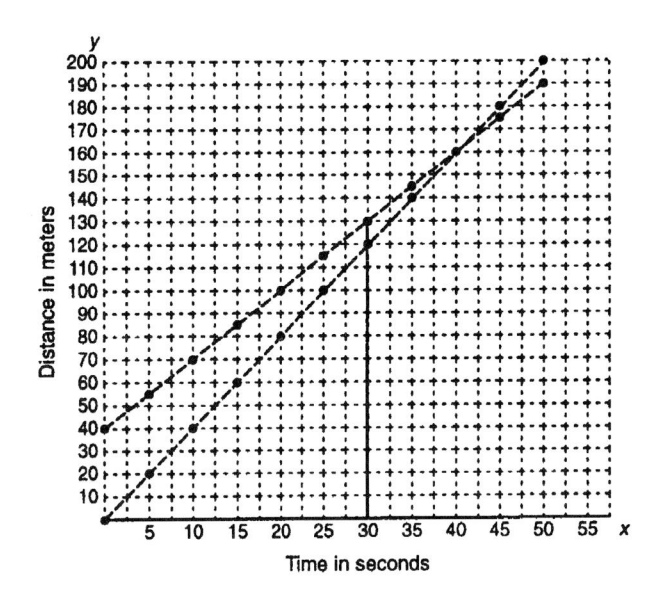

c. The graph shows the lines crossing after 40 seconds. At that time they will both be 160 meters from the swing set.

d. Pat will win the race. She catches up to Hal at the 40 second point and passes him then because she is going faster.

e. On the graph we can see that Pat reaches the soccer goal net after 50 seconds. At 4 meters per second she can go 200 meters in 50 seconds. After 50 seconds Hal has gone 50 × 3 = 150 meters, which added to his 40 meter head start puts him at the 190 meter point. This is 10 meters away from the soccer goal net.

33. a. There are about 115 small spaces on the graph. At .04 seconds per space this represents a time period of about 115(.04) = 4.6 seconds.

b. The signal lasted for a distance of 3 small spaces on the graph. This represents a time span of 3(.04) = .12 seconds.

c. There are about 16 or 17 spaces from the end of one pulse to the end of the next one. This is a time of about 17(.04) = .68 seconds. There is more than one pulse per second, so we know the pulse rate is faster than 60 beats per minute. To find the rate we can divide 60 seconds per minute by .68 seconds per beat to get 88 beats per minute. (Counting 16 spaces per pulse instead of 17 gives a rate of 94 beats per minute.)

35. a. Each new figure in the pattern increases the "steps going up" by one, the "steps going down" by one, and the number of tiles on top by 2. There are four additional tiles in each new figure. The table below represents the first 8 figures in the pattern.

| Figure Number | 1 | 2 | 3 | 4 | 5 | 6 | 7 | 8 |
|---|---|---|---|---|---|---|---|---|
| f(n) | 3 | 7 | 11 | 15 | 19 | 23 | 27 | 31 |

b. The graph shows these pairs of numbers plotted as coordinates of points. From the graph we can see that the points lie on a line. As the figure number increases by one, the number of tiles increases by four.

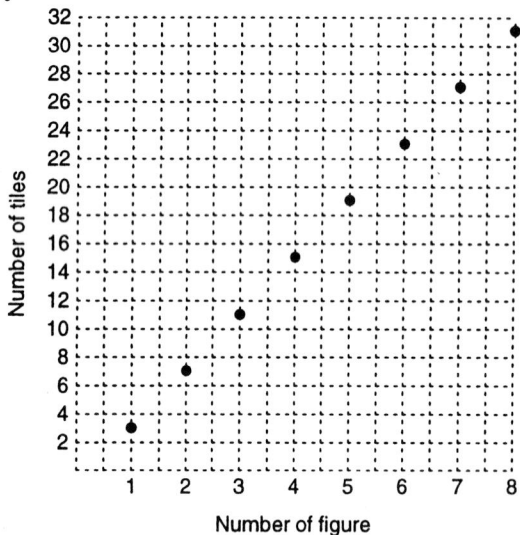

c. There are several different ways to describe the pattern. One is given in the part a solution. Here is another. The nth figure has n steps going up, n steps going down, and $2n - 1$ tiles on the top. Using this description, the 20th figure has $20 + 20 + 39 = 79$ tiles. So $f(20) = 79$.

d. Using the description from part c $f(n) = n + n + 2n - 1$. This is equivalent to $f(n) = 4n - 1$. There are other equivalent ways to express this rule.
$f(350) = 4(350) - 1 = 1399$.

37. a. In Sequence 1 each figure has a block of 20 tiles followed by a stack of an odd number of tiles. The number of tiles in the stack is twice the figure number minus 1. The number of tiles in the nth figure in Sequence 1 is $20 + 2n - 1$. Or, equivalently, it is $2n + 19$.
The nth figure in Sequence 2 can be viewed a number of ways. One way is to see n tiles in the middle of the top sandwiched between two stacks of $2n - 1$. This would give an algebraic rule of $n + 2n - 1$. Another way is to see a stack of n tiles on each side topped by a row of $n + 2$ tiles. This way gives an expression of $n + n + n + 2$. A third viewpoint sees each figure as a rectangle with a smaller square removed from it. Then the nth figure consists of an $n + 1$ by $n + 2$ rectangle with an n by n square removed. Algebraically this is $(n + 1)(n + 2) - n^2$. All of these expressions for Sequence 2 are equivalent to $3n + 2$.

b.

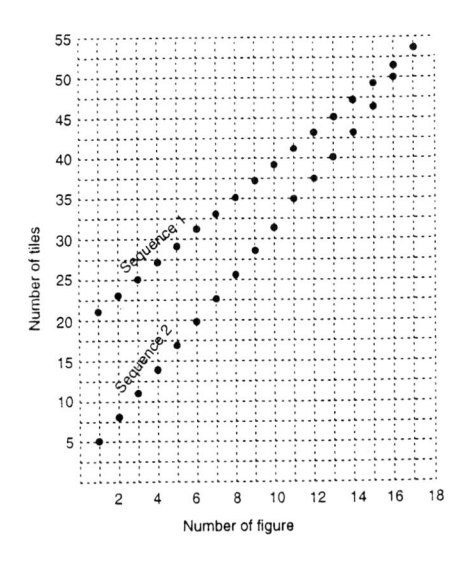

c. The graphs intersect when n has a value of 17.

d. The point of intersection is (17,53). This tells us that the 17$^{th}$ figure in each sequence has 53 tiles. Since (17,53) is the only intersection point, 17 is the only figure number in which both sequences have the same number of tiles.

# Section 2.3

1. The second statement is the contrapositive of the first statement. They are logically equivalent. If one of the statements is true then the other one must be true also. The diagram shows that whenever Susan wears her hat it is windy. The windy circle is completely contained within the wearing a hat circle. If Susan does not wear her hat then we are outside of the larger circle, so we must also be outside the inner circle.

3. a. If one takes a hard line with a bill collector, then it may lead to a lawsuit.

   b. If one is an employee of Tripak Company, then one must retire by age 65.

   c. If the class is to meet only twice a week, then there must be two-hour class sessions.

   d. If one is a pilot, then one must have a physical examination every 6 months.

5. a.    b.

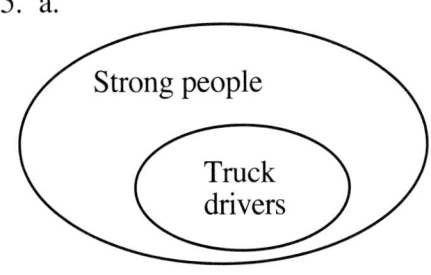

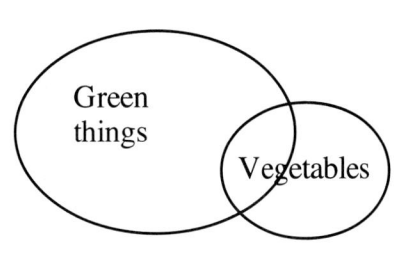

c.

```
┌─────────────────┐
│                 │
│  Animals with   │
│   two legs      │
│                 │
│    ┌──────┐     │
│    │ Ducks│     │
│    └──────┘     │
└─────────────────┘
```

d.

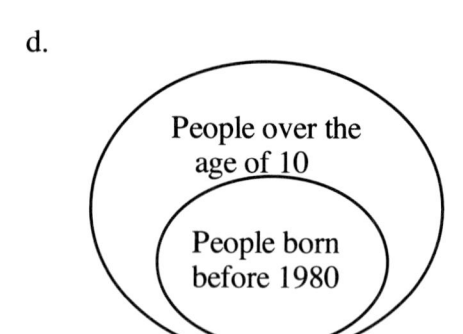

```
┌─────────────────┐
│  People over the│
│   age of 10     │
│                 │
│  ┌───────────┐  │
│  │People born│  │
│  │before 1980│  │
│  └───────────┘  │
└─────────────────┘
```

7. **Converse:** If you must itemize your deductions, then you take a deduction for your office.
   **Inverse:** If you do not take a deduction for your home office, then you do not have to itemize your deductions.
   **Contrapositive:** If you do not have to itemize your deductions, then you do not take a deduction for your home office.

9. **Converse:** If the camera focus is on manual, then switch B is pressed.
   **Inverse:** If switch B is not pressed, then the camera focus is not on manual.
   **Contrapositive:** If the camera focus is not on manual, then switch B is not pressed.

11. a. This is the converse of the statement. It is not logically equivalent to the statement. In general the converse may or may not be true. In this case the statement is false because sanctions may lead to agreement to inspections.

    b. This is the inverse of the statement. It is not logically equivalent to the statement.

    c. This is the contrapositive of the statement. It is logically equivalent to the statement.

13. The contrapositives are formed by converting the statement "If p then q" to the statement "If not q then not p".

    a. If the computer did not reject your income tax return, then you did not subtract $750 for each dependent.

    b. If the cards do not need to be dealt again, then there was an opening bid.

    c. If you do not return the books at the end of the week's free sing-a-long, then you were delighted with them.

15. A biconditional statement contains the phrase "if and only if" between the two parts of the statement. It means that the statement and its converse are both true.
    You pay the Durham poll tax if and only if you are 18 or older. (This statement may equivalently be stated as: You are 18 or older if and only if you pay the Durham poll tax.)

17. The two statements are: "If Robinson will be hired, then she meets the conditions set by the board" and "If she meets the conditions set by the board then Robinson will be hired".

19. Since all roses are flowers the set of roses is a subset of the set of flowers. Similarly, the set of flowers is a subset of the set of beautiful things. The conclusion is valid because the set of roses is completely contained in the set of beautiful things.

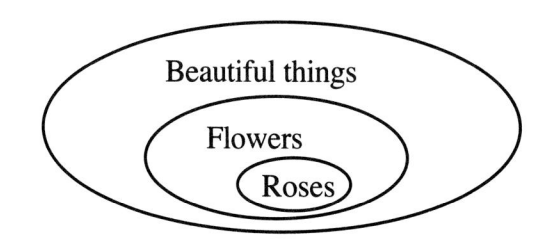

21. 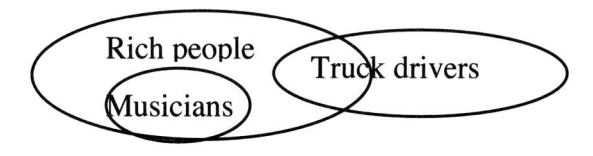 This Venn diagram shows why this is invalid reasoning. It may be that none of the rich truck drivers are musicians. All musicians are rich does not imply that all rich people are musicians.

23. The conclusion "This patient does not have anemia" is valid by the Law of Contraposition. The Venn diagram shows why.

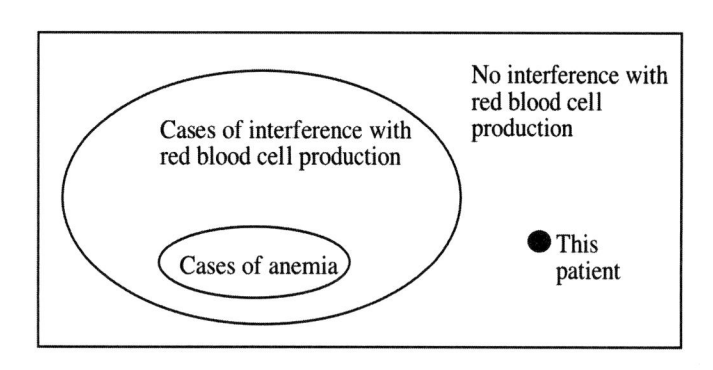

25. The conclusion "Mr. Keene has sufficient vitamin K in his body" is valid by the Law of Contraposition. The Venn diagram below shows why.

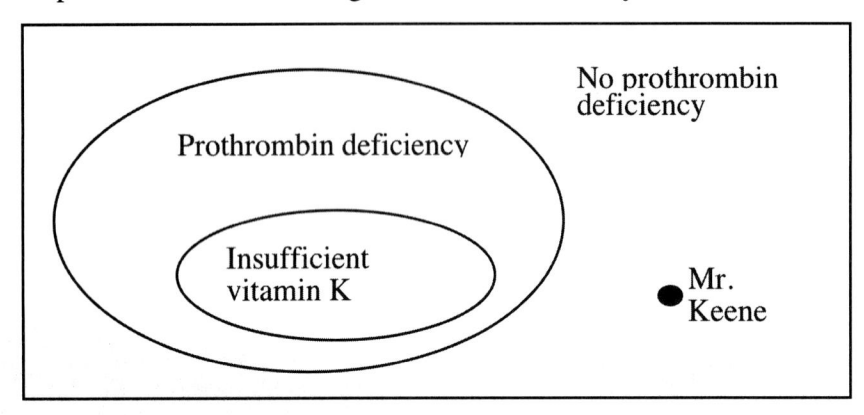

27. The conclusion is invalid. You could use a hexracket but not be a great tennis player.

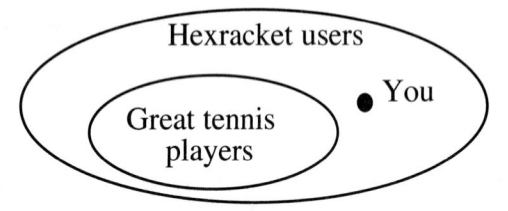

29. The conclusion is invalid. You could have longer rides without using the new clubs, as the Venn diagram shows.

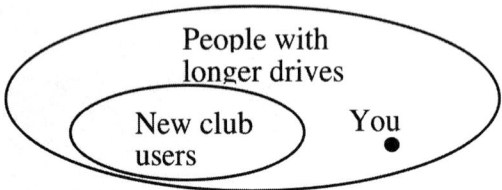

31. The conclusion is valid. All those who take Sleepwell have extra energy. If you are not in the set of those with extra energy, you can not be in the set of those taking Sleepwell.

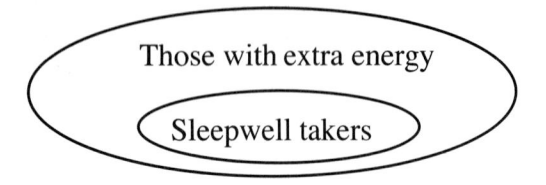

## 2.3 Introduction to Deductive Reasoning

33. One approach to this problem is to make a table with names along one edge and occupations along another. In this problem some of the clues also deal with gender, so we will also indicate genders that match with names and occupations.

| Gender | | M Appraiser | M Broker | F Cook | F Painter | M Singer |
|---|---|---|---|---|---|---|
| F | Dow | no | no | | | no |
| M | Elliot | | | no | no | |
| M | Finley | | | no | no | no |
| M | Grant | no | | no | no | no |
| F | Hanley | no | no | no | | no |

From clue 1 and 2 we know that the appraiser, broker, and singer are all male. From clue 3 Dow and Hanley are female. Since we know three are men, we get the rest of the gender information. Then the indicated slots can be marked no either because of gender mismatch or from clues 2, 4, and 5. So Elliot must be the singer, Hanley must be the painter, and Dow must be the cook. This lets us fill in some more no slots. After that we can see that Finley is the appraiser and Grant is the broker.

35. The conclusion is not valid. The way to refute the conclusion is as follows. By the contrapositive of statement 2, it is easy, so Huiru does not have trouble. Then by the contrapositive of 4, Huiru does not get dizzy. Then by statement 3 Huiru understands the proof. Finally, by the contrapositive of statement 1, the proof is arranged in logical order.

# Chapter 2 Test

1. a. The syh is the only piece that is both hexagonal **and** yellow.

   b. The sbt, sbr, sbh, and syt are all either triangular **or** black.

   c. The sbr, syt, syr, and syh are in the set of yellow **or** rectangular pieces. We are looking for the complement of this set. The answer is the two remaining pieces: sbt and sbh. They are neither yellow nor rectangular.

2. a. The intersection of A and B consists only of elements that are in both. This is {2, 4}.

   b. The union of A and B consists of all elements that are in either one or the other. This is {1, 2, 3, 4, 6}.

   c. The complement of A is {0, 1, 3, 5}. The intersection of this set with B is {1,3}.

   d. The complement of B is {0, 5, 6}. The union of this set with A is {0, 2, 4, 5, 6}.

3. a. b. 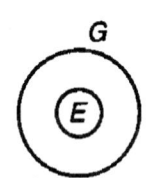 c.

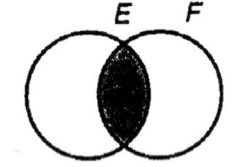

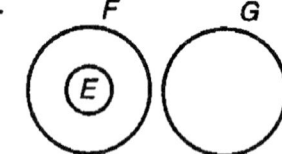

   d. 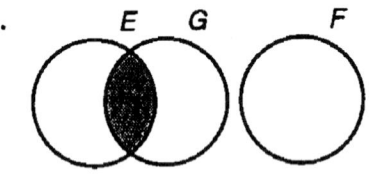 e. 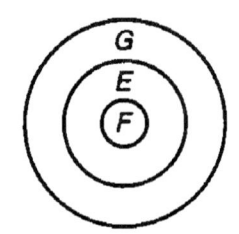 f.

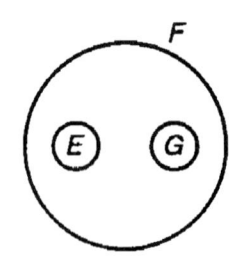

4. a. All of set A is shaded except for the part that is also in set B. This could be called A ∩ B'.

   b. Everything is shaded except the intersection of A and B. We can label this as (A ∩ B)'. It also could be viewed as the union of the complements of A and B. This is A' ∪ B'.

5.  a. The element k can be in R ∪ S without being in the intersection of R and S.

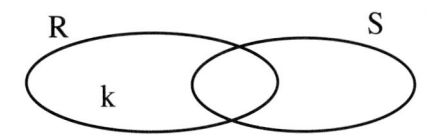

    b. Yes, if an element is in the intersection of two sets, then it is also in the union. The intersection is a subset of the union.

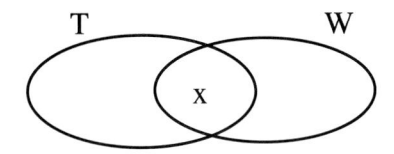

    c. No, if the element y is in the intersection of R and S, then it is contained in both R and S. Since it is in S it is not in the complement of S.

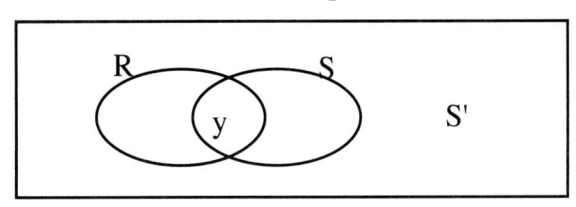

6.  a. This function takes as input the length of a line segment. The output is then half of this length. The range values will be half of the values in the domain. Since the domain is the numbers 1, 2, 3, 4, and 5, the corresponding range is .5, 1, 1.5, 2, 2.5.

    b.

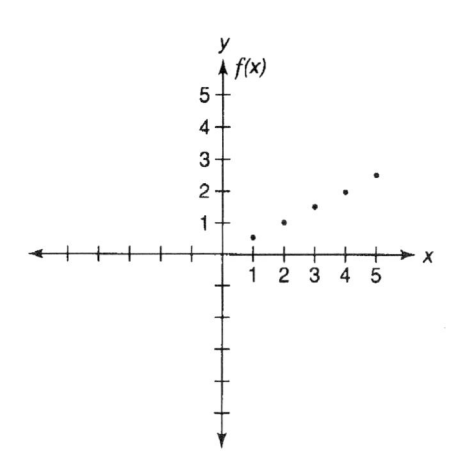

    c. An equation for this function is $f(x) = .5x$

    d. This function is linear. We can tell it is linear because its graph is a line.

7.  a. Between the two marked points the rise is –2 and the run is 4. The slope is –2/4 or -0.5.

    b. Between the two marked points the rise is 2 and the run is 4. The slope is 2/4 or 0.5.

8. To find the slope of the line we subtract 13 – 4 to get a rise of 9 and subtract 3 – 0 to get a run of 3. So the slope is 9/3, or 3. We also know the y-intercept because we are given the point (0,4). So we can write the equation y = 3x + 4 to represent the line through these two points.

9. a. To find the total cost we multiply $55 by the number of days and add this to the $120 charge for insurance. So the equation y = 55x + 120 gives the total cost of renting the trailer.

   b. Substituting 10 for x in the equation, we see that the cost of renting the trailer for 10 days is 55(10) + 120 = 550 + 120 = $670.

   c. To find the number of days you can rent the trailer within a budget of $850, we are looking for the largest value of x that will keep the value of 55x + 120 from going over 850. We can find this by solving the equation 55x + 120 = 850. Solving this equation gives a value of x between 13 and 14. So you can rent the trailer for a maximum of 13 days.

10. a. Since x represents the number of holiday hours worked and the holiday rate is $24 per hour, the expression 24x gives the total amount of money received for the holiday hours worked.

    b. We know that a total of 60 hours were worked, so if x is the number of holiday hours, then 60 – x is the number of non-holiday hours.

    c. Since 60 – x represents the number of non-holiday hours worked and the non-holiday rate is $15 per hour, the expression 15(60 – x) gives the total amount of money received for the holiday hours worked.

    d. If we add the amounts of money from the expressions in parts a and c we should get the total pay of $1062. So to find the holiday hours we can solve the equation $24x + 15(60 - x) = 1062$. Using the distributive property and then simplifying the left side by combining like terms, we get the equation $9x + 900 = 1062$. Solving this equation gives x = 18. So the electrician worked 18 holiday hours.

11. On Monday and Thursday Alanna stopped in the middle of her journey. This is indicated by the horizontal sections on those graphs, where the distance doesn't increase over time. Monday's stop was shorter, so Monday she jogged. Thursday she skateboarded. Wednesday she biked and Tuesday she walked. Her time for covering the same distance was shorter on Wednesday than it was on Tuesday.

12. One approach is to draw a Venn diagram. Numbers in the diagram were determined as follows. There were a total of 75 cars, so there will be a total of 25 in regions a, b, and c. The two circles contain 18+12 = 30 cars, so there must be 30 − 25 = 5 cars in the intersection. The amounts in regions a and c then follow. Five cars need both repairs.

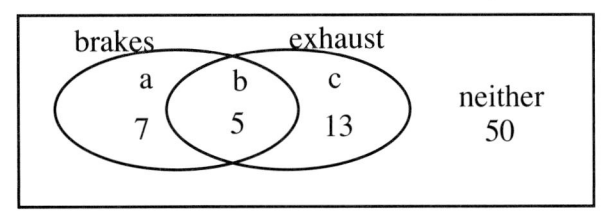

```
┌─────────────────────────────────────┐
│  brakes        exhaust              │
│    ⎛  a ⎞⎛ b ⎞ c                    │
│    ⎝  7 ⎠⎝ 5 ⎠13      neither       │
│                          50         │
└─────────────────────────────────────┘
```

13. Again a Venn diagram is a useful approach to this problem. We can start by putting 30 men in region e, those who are married and have a phone and a car. Then regions d, b, and f can be calculated by subtracting 30 from the information on these intersections. Next we can find the amounts in regions a, c, and g. Adding all of the numbers found so far gives 130. Since there are 150 men in the town, region h contains 20 men. These are the single men with no telephone and no car.

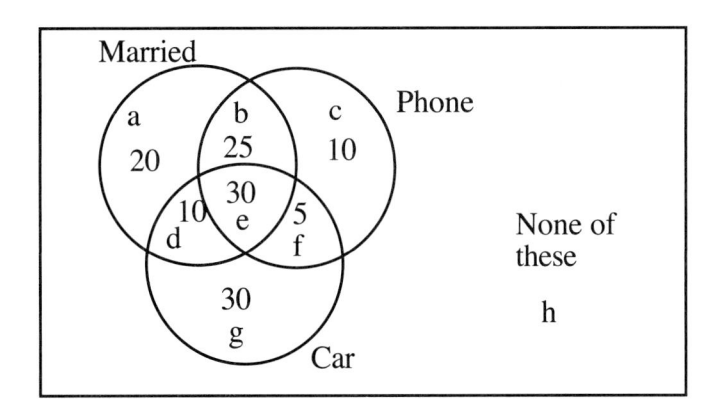

```
┌─────────────────────────────────────┐
│ Married                             │
│    ⎛ a  ⎞⎛ b ⎞⎛ c ⎞   Phone         │
│    ⎝ 20 ⎠⎝25 ⎠⎝10 ⎠                 │
│         ⎛ 30 ⎞                      │
│       10⎛ e ⎞ 5        None of      │
│         d    f         these        │
│          ⎛ 30 ⎞                       h │
│          ⎝ g  ⎠                     │
│            Car                      │
└─────────────────────────────────────┘
```

14. a. If people are denied credit, then they have a right to protest to the credit bureau.

   b. If a child was absent yesterday, then that child was absent again today.

   c. If a person came to the party, then that person received a gift.

15. a. **Converse:** If her husband goes with her, then Mary goes fishing.
      **Inverse:** If Mary does not go fishing, then her husband does not go with her.
      **Contrapositive:** If her husband does not go with her, then Mary does not go
                    fishing.

    b. **Converse:** If you receive 5 free books, then you join the book club.
       **Inverse:** If you don't join the book club, then you won't receive 5 free books.
       **Contrapositive:** If you don't receive 5 free books, then you didn't join the book
                     club.

16. The statement is logically equivalent to its contrapositive. This is statement 3.

17. "The prisoners will be set free if and only if there are peace talks". This can also be stated equivalently as "There will be peace talks if and only if the prisoners will be set free".

18. a. The conclusion is invalid. Not all aggresive birds are mallards. The black ducks that are aggresive birds may not be included in the subset that is the mallards.

    b. The conclusion is valid as shown in the Venn diagram below. (It would seem that geometry classes should be a subset of math classes, although that is not stated in the premises. Even so, it is stated that some math classes are geometry classes and all of these are interesting. Could all math classes be interesting?!)

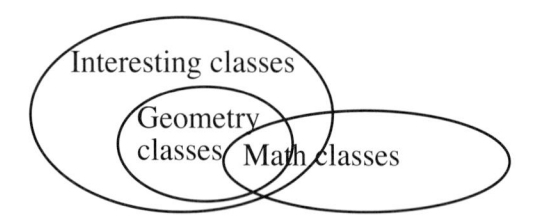

    c. The conclusion is invalid as shown in the Venn diagram.

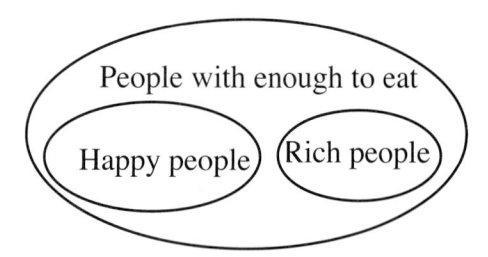

19. a. The contrapositive of the first premise is logically equivalent. It says, If a person does not have 10 times as much lung tissue as necessary then that person is not healthy. Since the people in ward B have less lung tissue than necessary, we can conclude that the people in ward B are not healthy.

    b. Since the hypothesis is true, the conclusion must follow. The game pieces should be set up as they were before the move. This is the Law of Detachment.

20. a. The Venn diagram below shows that the conclusion is invalid. The key word is **may**. If the statement had said you **will** keep the books, then the conclusion would have been valid.

    b. The conclusion is valid by the Law of Detachment.

## Chapter 3 Whole Numbers
## Section 3.1

1. The left side of the reckoning table has 1 thousand, 2 hundreds, 4 tens, and 1 unit. This is $1000 + 200 + 40 + 1 = 1241$.

3. This is an additive system, with two names only, Neecha for 1 and Boolla for 2. Our number 4 is two twos, so it is Boolla Boolla. The name for 5 would be Boolla Boolla Neecha because 5 is $2 + 2 + 1$. Similarly, $6 = 2 + 2 + 2$, so 6 is Boolla Boolla Boolla.

5. a. A hand has a value of 5. The number 22 can be thought of as four 5s and 2 ones. The name for 22 in this system is 4 hands and 2.

   b. If 25 is a "hand of hands", then think of 37 as $25 + 10 + 2$. The name for 37 is one hand of hands, 2 hands, and 2.

7. The set here has 19 longs and 24 units. The 24 units could be traded for 2 longs and 4 units. Then there are 21 longs and 4 units. Twenty of the 21 longs could be traded for 2 flats. The minimal collection is 2 flats, 1 long, and 4 units.

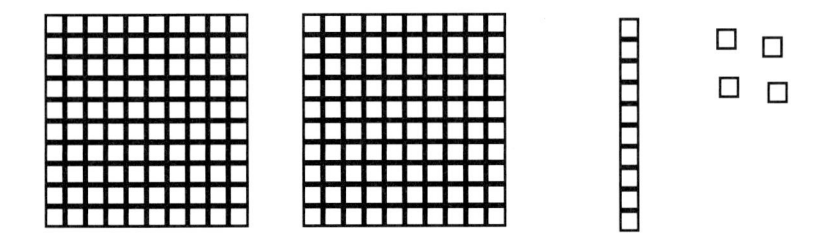

9. a. The base seven long consists of 7 units. The flat is a square of $7 \times 7 = 49$ units.

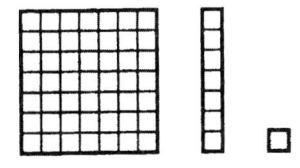

   b. In base 3 a long is 3 units and a flat is $3 \times 3 = 9$ units.

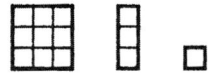

11. The pictured set has 72 units. In base seven flats have 7 × 7 = 49 units and longs have 7 units. Since 72 = 49 + 3 × 7 + 2, the units in the set can be traded in for 1 flat, 3 longs, and 2 units in base 7. The notation is 132$_{seven}$. In base five flats have 5 × 5 = 25 units and longs have 5 units. Since 72 = 2 × 25 + 4 × 5 + 2, the units in the set can be traded in for 2 flats, 4 longs, and 2 units. The notation is 242$_{five}$. See the diagrams below.

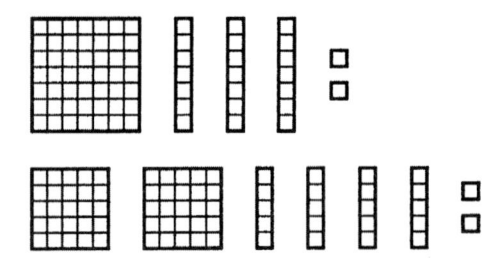

13. a. In base five, flats are 25 units and longs are five units. 3 × 25 + 4 × 5 + 3 = 98 units.

    b. In base eight, flats are 64 units and longs are 8 units. 6 × 64 + 0 × 8 + 5 = 389 units.

15. In each numeration system, the symbol for 1 is repeated to create the symbols for 2, 3, and 4. In the Babylonian and Egyptian systems, the symbol for 1 is repeated in the symbols for 2 through 9. Grouping by 5s occurs in the Roman and Mayan systems. In these systems, a symbol for 5 is used with the symbols for 1, 2, 3, and 4 to form the symbols for 6, 7, 8, and 9; the symbol for 10 can be formed by combining two symbols for the number 5. The Egyptian and Babylonian systems appear to be the most solidly based on ten, since that is the first new symbol. The Mayan system is the one most solidly based on five.

17. a. In Egyptian numeration we simply draw the amount of the appropriate symbol for each place value. For 40,208 that is four 10,000's, two hundreds, and eight ones.

    b. Roman numeration also has symbols for 5, 50, 500, etc. We think of the number 1776 as one 1000, one 500, two 100's, one 50, two 10's, one 5, and one 1.
    This is M D C C L X X V I.

    c. The Babylonian system is base on 60. Instead of 1's, 10's, and 100's as place values, now we are working with 1's, 60's, and 3600's. To write the number 4635, we think of it as 3600 + 1035, which is 1 × 3600 + 17 × 60 + 15 × 1. When a position had a value of more than 9, the Babylonians used a different symbol to show 10.

d. The Mayan system is based on 20's. In 172 there are eight 20's plus 12 ones. The symbol for 8 is three dots above one line, representing 3 ones and a five. The symbol for 12 is two dots above two lines, representing 2 ones and 2 fives. The 8 is placed above the 12 to indicate 8 twenties and 12 ones.

19. This is a modified base 10 system similar to the Roman numeration system. There are symbols for 1, 5, 10, 50, 100, 500, etc.

| Hindu-Arabic | 1 | 4 | 8 | 16 | 26 |
|---|---|---|---|---|---|
| Attic-Greek | I | IIII | ΓIII | ΔΓI | ΔΔΓI |

| Hindu-Arabic | 32 | 52 | 57 | 206 | 511 |
|---|---|---|---|---|---|
| Attic-Greek | ΔΔΔII | ΡII | ΡΓII | HHΓI | ΡΔI |

21. We can write expanded form as a sum with the symbol from each place being multiplied by its place value. Note that this is indeed the number. The place values can either be written as powers of 10 or in longer form. (The zero terms are optional.)

  a. $7,082,555 = 7 \times 10^6 + 0 \times 10^5 + 8 \times 10^4 + 2 \times 10^3 + 5 \times 10^2 + 5 \times 10 + 5$
  or $7 \times 1,000,000 + 0 \times 100,000 + 8 \times 10,000 + 2 \times 1000 + 5 \times 100 + 5 \times 10 + 5$

  b. $57,020 = 5 \times 10^4 + 7 \times 10^3 + 0 \times 10^2 + 2 \times 10 + 0$
  or $5 \times 10,000 + 7 \times 1000 + 0 \times 100 + 2 \times 10 + 0$

23. a. In 1478, the number is one thousand, four hundred seventy eight. The four is in the hundreds place and represents 400.

  b. The underlined digit is in the thousands place. Its value is zero.

  c. The two is in the millions place. Its value is 2,000,000.

25. a. Four thousand forty

  b. Seven hundred ninety-three million, four hundred twenty-eight thousand, five hundred eleven.

  c. Thirty million, one hundred ninety-seven thousand, seven hundred thirty-three.

  d. Five billion, two hundred ten million, nine hundred ninety-nine thousand, six hundred seventeen.

27. a. In the number 43,668,926 there are 436 hundred thousands plus more than 50,000. The number is closer to 43,700,000 than 43,600,000, so we round it to 43,700,000.

   b. It is closer to 43,670,000 than 43,660,000, so we round it to 43,670,000.

   c. It is closer to 43,669,000 than 43,668,000 so we round it to 43,669,000.

   d. It is closer to 43,668,900 than 43,669,000 so we round it to 43,668,900.

29. Below are sketches of the various models of the numbers.

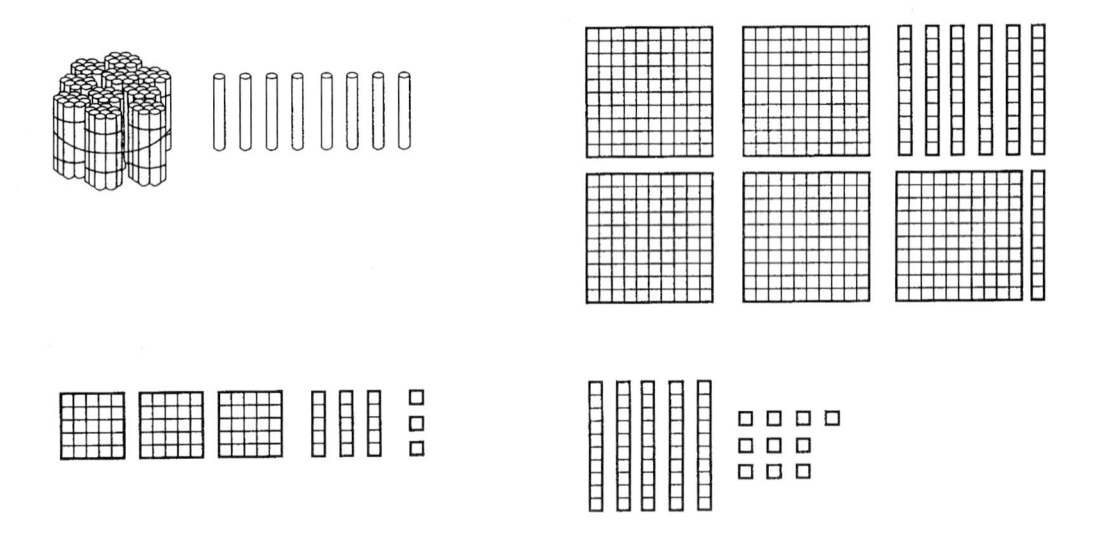

31. a. The only difference between the digits of 72,913,086 and 78,913,086 is found in the millions place. If we add 6,000,000 to the first number we will get the second. This will not change the other digits.

   b. Here the digits are different in the 1000's place and in the 100's place. Adding 200,300 to the first number will give the second one. Another way would be to use the two steps of adding 200,000 and adding 300 in either order.

   c. This one is trickier. It doesn't look like there is much different between 7,496,146 and 749,146, but all but the smallest three place values have been changed. One solution would be to subtract 6146 from 7,496,146 giving 7,490,000. Then dividing this number by 10 gives 749,000. Finally add 146 to get 749,146. Another different solution is given in the back of the text.

33. a. We are looking for the smallest 9 digit whole number that contains each of the 9 digits once. The digit with the largest value will be the first one, in this case in the 100,000,000's place, so we place the 1 there. Similarly, we need the 2 next, etc. The number is 123,456,789.

b. Here we are allowed to repeat digits. Our smallest available digit is 1, so we will use all 1's. The number is 111,111,111.

35. a. Multiplication is done before addition. The calculator reads the expression as 700,000 + 7,000 + 7. The result displayed is 707007.

b. Again multiplication is done first. This gives 12000 + 800 + 30 + 2. The calculator will display 12832.

37. Here are a few more examples of numbers that can be written as sums of binary numbers:

$$9 = 8 + 1 = 2^3 + 2^0$$
$$30 = 16 + 8 + 4 + 2 = 2^4 + 2^3 + 2^2 + 2^1$$
$$76 = 64 + 8 + 4 = 2^6 + 2^3 + 2^2$$

A reasonable conjecture to make is: "All whole numbers are either binary numbers or can be written as sums of binary numbers." *[Do you think that this statement is true? Can you prove or disprove it?]*

39. If the digits are consecutive numbers, we are looking for a number such as 123 or 567 or 789. However, none of these choices satisfies the second condition, that the sum of of each pair of digits is greater than 4 and less than 10. In 123 1 + 2 is not greater than 4. In 567 and 789 there are pairs of digits whose sum is more than 10. According to the third condition, the middle digit must be even. This means we can choose from 123, 345, 567, and 789. Our only candidate is 345. It does satisfy all of the conditions. One could also argue that 543 is a correct answer, depending on the interpretation of the word "consecutive".

41. We need to find how many times the digit 3 is used in writing all of the whole numbers from 300 to 599 inclusive. A 3 is used in the 100's place of all 100 of the numbers from 300 to 399. So we will need 100 bronze 3's for the 100's place digits. In the 10's place there will be 3's in 330-339, 430-439, and 530-539. This is another 30 3's. In the 1's place there will be 3's in 303, 313, 323, ... 583, 593. There are 30 more 3's needed here (10 in each 100). A total of 100 + 30 + 30 = 160 3's are needed.

## Section 3.2

1. a. The units wheel would make more than one revolution because it is turned to the 4 and then 9 more units. This is more than 10. The 100's wheel would also make more than one complete revolution.

   b. The machine should give the same result no matter what order we enter the numbers.

3. a. We are working in base five, so any time there are more than four of any type of piece we will trade five of them for one of the next size larger. Combining the sets we have 3 flats, 5 longs, and 5 units. The 5 units can be traded for one long. Five longs can be traded for 1 flat. The result is 4 flats, 1 long, and no units. Positional numeration for these sets is written this way: for set A: $232_{five}$ for set B: $123_{five}$ for the sum: $410_{five}$.

   b. We are working in base twelve, so any time there are more than eleven of any type of piece we will trade twelve of them for one of the next size larger. Combining the sets we have 10 flats, 14 longs, and 7 units. Twelve of the 14 longs can be traded for one more flat. The result is 11 flats, 2 longs, and 7 units. Positional numeration for these sets is written this way: for set A: $852_{twelve}$ for set B: $295_{twelve}$ for the sum: $E27_{twelve}$. Since we need to represent 11 flats, we need the new symbol E.

5. a. We are working in base eight, so any time there are more than seven of any type of piece we will trade eight of them for one of the next size larger. We are starting with set B, which contains 2 flats, 6 longs, and 5 units. We want to add a set of pieces to this set to end up with set A: 5 flats, 2 longs, and 3 units. Adding 6 units to the 5 units will give us 3 units and an extra long. Now we have 7 longs. We want 2, so we will add 3 to 7. This gives us 2 longs and an extra flat. Now we have 3 flats. We need 5 so we will add 2 more. The amount we added in was $236_{eight}$. We found the difference $523_{eight} - 265_{eight} = 236_{eight}$.

   b. We are working in base five, so any time there are more than four of any type of piece we will trade five of them for one of the next size larger. We are starting with set B which contains 1 flats, 3 longs, and 4 units. We want to add a set of pieces to this set to end up with set A: 3 flats, 4 longs, and 2 units. Adding 3 units to the 4 units will give us 2 units and an extra long. Now we have 4 longs which is what we want. We also need to add 2 more flats. The amount we added in was $203_{five}$. We found the difference $342_{five} - 134_{five} = 203_{five}$.

7. See the diagrams on the next page. In part a the sum contained 14 units, so 10 of them were traded for one long. In part b we needed to take 3 units away, but there was only one unit, so one long was traded for five more units. In part c we are also doing subtraction, but we want to use the comparison concept instead of take-away. So the diagram shows both 161 and 127. We trade one of the longs in 161 for 10 units so that we can see the difference between the units. In part d regrouping is done twice. Six units are traded for one long and one unit. Five longs are also traded for a flat.

## 3.2 Addition and Subtraction

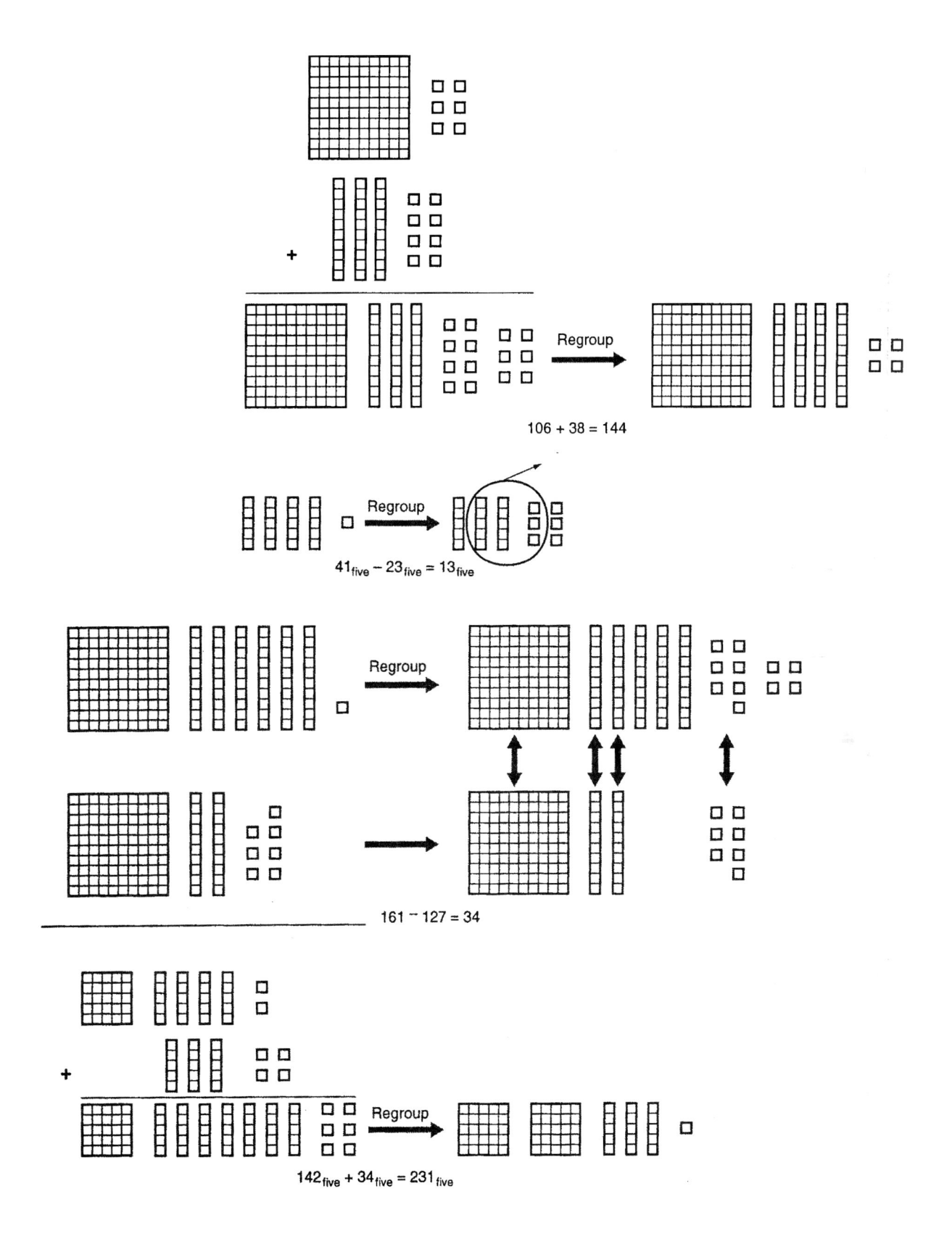

$106 + 38 = 144$

$41_{\text{five}} - 23_{\text{five}} = 13_{\text{five}}$

$161 - 127 = 34$

$142_{\text{five}} + 34_{\text{five}} = 231_{\text{five}}$

9. See the diagrams of the number lines below. In part a 2 + 5 is shown by the arrows below and 5 + 2 by the arrows above. In part b the arrows below show 1 + (2 + 4) and the arrows above show (2 + 4) + 1.

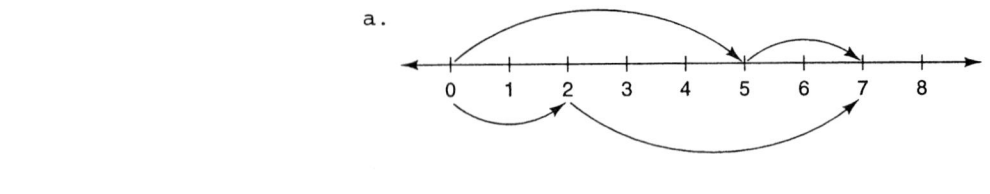

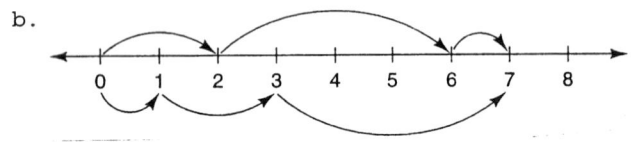

   a.

   b.

11. In part a we first draw an arrow from 0 to 6 then another arrow back to 3 to indicate 6 – 3, then a third arrow from 3 to 1 indicating a further subtraction of 2. (6 – 3) – 2 = 1.
    In part b an arrow goes from 0 to 6 and then another arrow back to 0, showing 6 – 6 = 0.

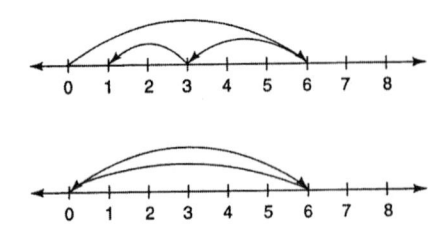

13. a. In left-to-right addition we start with the digits on the left. In this case add 7 hundreds and 5 hundreds to get 1 thousand and 2 hundreds. Then add the tens, 2 + 0 = 2 tens. Last add the units. Adding 6 + 8 gives 14 units, which is 1 ten and 4 units. The extra ten is added to the 2 tens to give 3 tens. The answer is 1234. One advantage of this method is that it is done in the same direction in which we read.

    b. Computing 974 + 382 by partial sums we get 6 units + 15 tens + 12 hundreds.
    Then we regroup the 15 tens into 1 hundred and 5 tens, and regroup the
    12 hundreds into 1 thousand and 2 hundreds.
    Combining, we have 1 thousand, 3 hundreds, 5 tens, and 6 units.
    This can be shown as is done here on the right.
    One advantage of this method is that the regrouping can be done in any order.

    ```
      974
    +382
        6
       15
       12
     1356
    ```

15. a. This is an example of the Associative property for addition because the grouping of the numbers has been changed. On the left side we associate the first two numbers together. On the right side we associate the last two numbers. The sum is the same on both sides.

    b. This is an example of the Commutative property of addition. Here the order of addition of 47 + 62 has been changed to 62 + 47.

17. a. Subtraction is not commutative. For example, $7 - 5 = 2$ but $5 - 7 = -2$.

   b. Addition is closed on the set of even numbers because whenever two even numbers are added the result is an even number. In other words, when the operation of addition is performed on elements in the set of even numbers the result is still an element of that set.

19. a. The student added the columns of numbers correctly, but there was a sum of 13 units, so there should be an additional ten in the answer. In other words, the student did not "carry" the extra ten.

   b. The error here was probably made when the student added 6 and 8 to get 14 units. Instead of recording 4 units and adding one more to the tens, it looks like this student recorded 1 unit and added 4 more to the tens.

21. a. The student made the error of subtracting $6 - 4$ instead of trading one of the tens so that the subtraction could be $14 - 6$.

   b. The error here was in forgetting that one of the 5 tens in 52 was needed to create additional units to subtract the 8 from.

23. a. $23 + 25 + 28 = 20 + 3 + 20 + 5 + 20 + 8 = 20 + 20 + 20 + 3 + 5 + 8 = 60 + 16 = 76$
   or
   $23 + 25 + 28 = 25 + 23 + (2 + 26) = 25 + 25 + 26 = 50 + 26 = 76$.
   Other methods also work.

   b. $128 - 15 + 27 - 50 = 128 + 27 - 15 - 50 = 155 - 65 = 90$
   or
   $128 - 15 + 27 - 50 = 128 + 12 - 50 = 140 - 50 = 90$.
   Other methods also work.

   c. $83 + 50 - 13 + 24 = 83 - 13 + 50 + 24 = 70 + 50 + 24 = 120 + 24 = 144$
   Other methods also work.

25. a. $6502 - 152 = 6500 - 150 = 6350$.   Here 2 was subtracted from both numbers first.

   b. $894 - 199 = 895 - 200 = 695$.   Here 1 was added to both numbers first.

   c. $14,200 - 2700 = 14,000 - 2500 = 11,500$.   200 was subtracted from both numbers first.

27. a. We want to find the difference 400 − 185. We add 15 to 185 to get to 200. Then add 200 to 200 to get 400. Since we added 15 and 200, the difference is 215.

b. We want to find the difference 535 − 250. We add 250 to 250 to get to 500. Then add 35 to 500 to get 535. Since we added 250 and 35, the difference is 285.

c. We want to find the difference 135 − 47. We add 53 to 47 to get to 100. Then add 35 to 100 to get 135. Since we added 53 and 35, the difference is 53+35 = 88. This problem could also be done in 3 steps by first adding 3, then 50.
(Other routes could be taken in each of these problems in #27)

29. a. Think 100 + 40 + 20 = 160.

b. Think 30 + 40 + 60 = 130.

31. a. 359 − 192 ≈ 360 − 190 = 170 or 359 − 192 ≈ 360 − 200 = 160.

b. 712 + 293 ≈ 700 + 290 = 990 or 712 + 293 ≈ 700 + 300 = 1000.

c. 882 + 245 ≈ 900 + 250 = 1150 or 882 + 245 ≈ 900 + 245 = 1145.

d. 1522 − 486 ≈ 1500 − 500 = 1000.

33. In front-end estimation we just look at the first digit of each number.

a. 362 + 408 + 978 ≈ 300 + 400 + 900 = 1600.

b. 16 + 49 + 87 + 33 ≈ 10 + 40 + 80 + 30 = 160.

c. 7215 + 5102 + 8730 ≈ 7000 + 5000 + 8000 = 20,000.

35. a. Rounding each bill to the nearest hundred we get
100 + 100 + 100 + 200 + 200 + 100 + 100 + 500 + 500 + 100 + 100 + 600 + 100 + 0.
This is a total of $2800.

b. The homeowner can not pay all of these bills using only a salary of $1800. Even if the rounding had all been in the overestimation direction, the maximum error would have been 14 × 50 = $700.

37. a. The sequence will be 8723, 8823, 8923, 9023, 9123, 9223, 9323, 9423. Each time the keystrokes + 100 = are repeated another 100 is added to the sequence.

b. The sequence is 906, 896, 886, 876, 866, 856, 846. Ten is subtracted each time.

## 3.2 Addition and Subtraction

39. a. The next 6 numbers are 2859, 3004, 3149, 3294, 3439, 3584.
    Each time the = key is pressed 145 is added to the previous number.

    b. The 10th number will be 4164. Since 145 is added 4 more times after 3584, it is
       $3584 + 145 \times 4 = 4164$.

    c. If 145 is subtracted each time instead of added, the numbers will be
       2569, 2424, 2279, 2134, 1989, 1844.

41. a. Pressing the keys 2 + 2 = = = = = will give a sequence starting with two
       and going forward by 2's because each time = is pressed another 2 is added.

    b. Pressing the keys 30 − 3 = = = = = will give a sequence starting with 30 and
       going backward by 3's because each time = is pressed another 3 is subtracted.

    c. Pressing the keys 20 + 5 = = = = = will give a sequence starting with 20 and
       going forward by 5's because each time = is pressed another 5 is added.

43. a. If all 22 of the cars with standard transmission also have air conditioning, then there are
       only 30 cars on the lot. This is the minimum because if any of the cars with standard
       transmission do not have AC, then there would be more cars.

    b. The maximum will occur when none of the cars have both features.
       In that case there are 52 cars.

    c. If 17 cars have both features, then there are 30 − 17 = 13 cars with only AC and
       22 − 17 = 5 cars with only standard transmission. This is a total of 17 + 13 + 5 = 35 cars.

    d. The case in part b is where the answer can be found by adding 22 + 30.

    Note: A Venn diagram may be useful in visualizing #43.

45. The Venn diagram shows that there were a total of
    11 + 15 + 6 = 32 students in the class.
    The 11 and 6 figures were given in the problem
    information. We can deduce the 15 by using the
    totals from the two days. No students were outside
    the sets because everyone watched on at least one of
    the days.

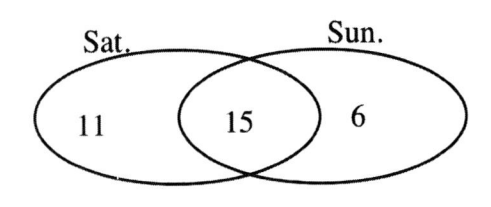

47. Pressing = will always subtract 83 from the previous number, so the six view screens will show
    these numbers:  a. 665   b. 724   c. 1143   d. 831   e. 1289   f. 572

49. The numbers can be paired as follows: 1 with 19, 2 with 18, 3 with 17, etc. This leaves only the number 10 without a partner. Each of the other pairs adds to 20. Place the 10 in the middle and the pairs of numbers opposite each other around the circle. Then three numbers on a line will always add to 30.

51. The value of C must be 9 because otherwise the sum would only be a 3 digit number. The value of G must be 1 because we are adding a number less than 10 to 999. Then the K is 2 because the sum ends in a 1. 999 + 2 = 1001, so the F represents 0. The sum is

    ```
      9 9 9
    +     2
    1 0 0 1
    ```

## Section 3.3

1. a. The upper 7 rows on the front face will each have their first 3 bulbs lit.

   b. The upper 2 rows on the front face will each have their first 8 bulbs lit.

3. a. The upper 6 rows on the front face will each have their first 4 bulbs lit. This is repeated in the next two arrays behind the front. 72 bulbs in the upper left corner will be lit.

   b. The top row on the front face will each have its first 8 bulbs lit. This will be repeated in the next seven arrays behind the front. Since the cube is 8x8x8, the whole top will be lit.

5. a. The sketch below shows three copies of the base 10 pieces for 168. Together the whole collection represents 168 × 3. To find the product, regroup the pieces. There are 24 units which can be regrouped to form 2 longs and 4 units. There are 6 × 3 = 18 longs, plus the extra two longs, a total of 20 longs. These can be traded for 2 flats. This makes a total of 5 flats, 0 longs, and 4 units. The product is 504.

   b. The diagram for 209 × 4 is similar. After sketching 4 copies of 2 flats, 0 longs, and 9 units we see a total of 8 flats and 36 units. Thirty of the units can be regrouped as 3 longs. Then we have 8 flats, 3 longs, and 6 units. The product is 836.

c. The diagram below shows $423_{\text{five}} \times 3$. Before regrouping there are 12 flats, 6 longs, and 9 units. In base five we trade 5 of the units for another long, trade 5 longs for another flat, and trade 10 flats for 2 long-flats. The product is $2324_{\text{five}}$.

d. Similarly, in multiplying $47_{\text{eight}} \times 5$ we look at 5 copies of $47_{\text{eight}}$. This would be $4 \times 5 = 20$ longs and $7 \times 5 = 35$ units. Of the 35 units, 32 could be regrouped as 4 more longs, leaving 24 longs and 3 units. The 24 longs could all be regrouped as 3 flats. The result is 3 flats, 0 longs, and 3 units. So $47_{\text{eight}} \times 5 = 303_{\text{eight}}$.

7. The arrow diagrams below show the products on the number line. In part c notice that 3 jumps of 4 units ends up at the same place as 4 jumps of 3 units.

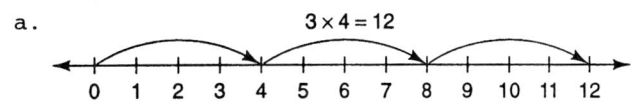

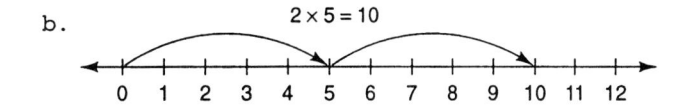

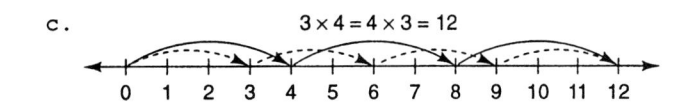

   a. $3 \times 4 = 12$

   b. $2 \times 5 = 10$

   c. $3 \times 4 = 4 \times 3 = 12$

9. a. It looks like the student multiplied $7 \times 4$ correctly and carried the 2 tens to the correct position, but then an error was made. Perhaps the student added the two 2's, or perhaps they were multiplied to get 4.

   b. Here again the first part was done correctly. Either the 2 which was carried was ignored, or the 2 and 1 in the ten's column were added to get 3. It would be productive to interview these students to see what they were thinking.

11. The diagrams below show area models for these two products. In part a note the rectangle with dimensions 7 × 4 and area 28 and the rectangle with dimensions 7 × 20 and area 140. In part b notice the four rectangles representing the four partial products.

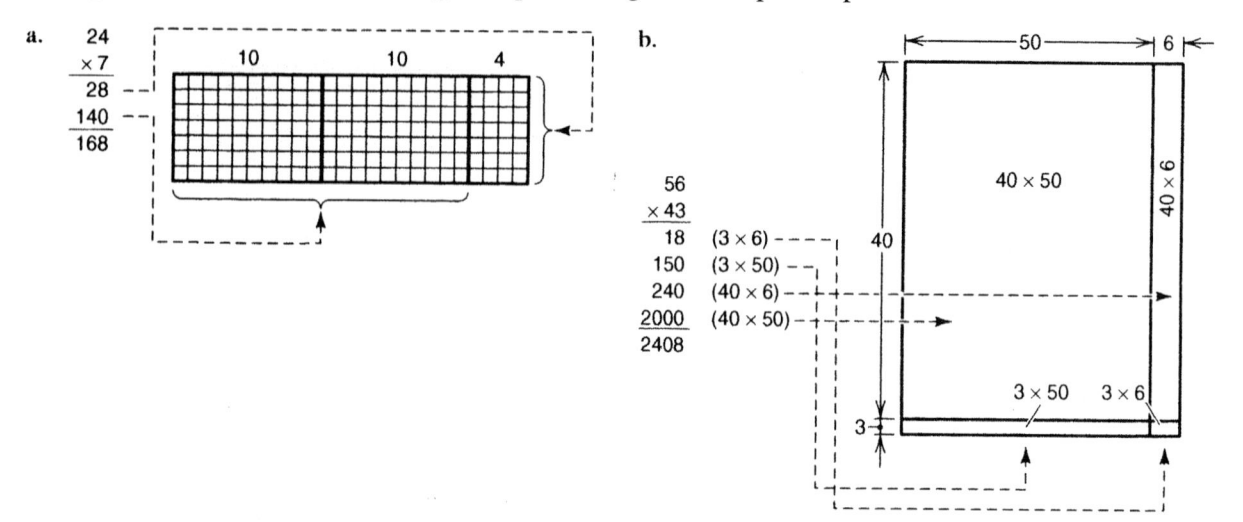

a.
```
  24
×  7
  28
 140
 168
```

b.
```
   56
 × 43
   18   (3 × 6)
  150   (3 × 50)
  240   (40 × 6)
 2000   (40 × 50)
 2408
```

13. a. The difference between the two expressions is that 2 × 7 on the left has been rewritten as 7 × 2 on the right. The equation is true by the Commutative property of multiplication.

    b. The difference between the two expressions is that (43 × 7) × 9 on the left has been rewritten as 43 × (7 × 9) on the right. The equation is true by the Associative property of multiplication.

    c. This is an example of the Distributive property of multiplication over addition. The quantity (12 + 17) can either be multiplied by the sum (16 + 5), or the (12 + 17) can be distributed so that it is first multiplied by 16, then by 5, and finally the two products are added.

15. a. If two odd whole numbers are multiplied, the result is always another odd whole number. Therefore the set of odd whole numbers is closed under multiplication.

    b. Addition is not closed on the set of whole numbers less than 100. For example, 90 + 80 is <u>not</u> an element of this set, but 90 and 80 <u>are</u> both elements.

    c. The set of whole numbers whose units digits are 6 includes elements such as 6, 16, 26, 36, 46, etc. Multiplying any two of these always gives another number ending in 6. Why? So this set is closed under multiplication.

## 3.3 Multiplication

17. a. We can use the commutative property to think of $2 \times 83 \times 50$ as $2 \times 50 \times 83$, which is $100 \times 83 = 8300$.

    b. In $5 \times 3 \times 2 \times 7$, the 5 and 2 are compatible. Multiplying these we get $10 \times 3 \times 7$. Then we can either think $10 \times 21$ or $30 \times 7$. Either way, the answer is 210.

19. a. To use the distributive property we either want to think of 25 as $20 + 5$, or think of 12 as $10 + 2$. Using the first method we get: $25 \times 12 = (20 + 5) \times 12 = 20 \times 12 + 5 \times 12 = 240 + 60 = 300$.
    The second method is: $25 \times (10 + 2) = 25 \times 10 + 25 \times 2 = 250 + 50 = 300$.

    b. Again there are two ways. Either $15 \times 106 = 15 \times (100 + 6) = 15 \times 100 + 15 \times 6 = 1500 + 90 = 1590$.
    Or, $(10 + 5) \times 106 = 10 \times 106 + 5 \times 106 = 1060 + 530 = 1590$.

21. a. Here we want to rewrite one of the factors as a difference of two numbers. We would like to choose a number that is just one or two less than a multiple of 10. We will rewrite 19 as $20 - 1$. $35 \times 19 = 35 \times (20 - 1) = 35 \times 20 - 35 \times 1 = 700 - 35 = 665$.

    b. We will rewrite 99 as $100 - 1$. $30 \times 99 = 30 \times (100 - 1) = 30 \times 100 - 30 \times 1 = 3000 - 30 = 2970$. We are thinking of 99 thirties as one thirty less than 100 thirties.

23. a. We want to find either a factor or a multiple of one of the numbers that is easy to multiply. In the product $24 \times 25$, we notice that $25 \times 4 = 100$, so it would be nice if we could multiply 25 by 4. We can do this if we also divide 24 by 4. The product $6 \times 100$ is equal to the product $24 \times 25$. One way to see this is to rewrite $24 \times 25$ as $(6 \times 4) \times 25 = 6 \times (4 \times 25)$. So $24 \times 25 = 6 \times 100 = 600$.

    b. The route is not so clear in the product $35 \times 60$. One way is to see 60 as $6 \times 10$. If we divide 60 by 6, then to keep an equal product we need to multiply 35 by 6. $35 \times 6 = 210$. So, $35 \times 60 = 210 \times 10 = 2100$. Another way is $35 \times 60 = 7 \times (5 \times 60) = 7 \times 300 = 2100$.

25. a. To estimate $22 \times 17$ we can round 22 to 20 and round 17 to 20. So our estimate is $20 \times 20 = 400$. Is this too high or too low? It should be a fairly close estimate since one number was rounded up and one was rounded down. It is likely to be a bit high because we rounded the 17 further up than we rounded the 22 down.

    b. To estimate $83 \times 31$ we round 83 to 80 and 31 to 30. The result is $80 \times 30 = 2400$. This estimate is definitely too low because both numbers were rounded down. An adjustment to improve the estimate might be to add another 80, since there were really 31 eighties, not 30 eighties; and add 3 more 30's. So $2400 + 80 + 90 = 2570$ would be a closer estimate. *[Can you see the exact answer from here?]*

27. a. Think of $4 \times 76 \times 24$ as close to $4 \times 25 \times 76$. This would be $100 \times 76 = 7600$. This estimate is too high because we replaced 24 with 25.

b. One estimate for 3 × 34 × 162 uses this line of reasoning: 3 × 33 = 99 and 3 × 34 is a little over 100, so the whole product is close to 100 × 162 = 16200. This estimate is too low because 3 × 34 is more than 100.

29. a. Using front-end estimation, 36 × 58 is approximately 30 × 50 = 1500. We get a closer estimate if we also multiply our rounded numbers by units digits.
  36 × 58 ≈ 30 × 50 + 30 × 8 + 50 × 6 = 1500 + 240 + 300 = 2040.

  b. Using front-end estimation, 42 × 27 is approximately 40 × 20 = 800. We get a closer estimate if we also multiply our rounded numbers by units digits.
  42 × 27 ≈ 40 × 20 + 40 × 7 + 20 × 2 = 800 + 280 + 40 = 1120.

31. a. An estimate for 18 × 62 would be 20 × 60 = 1200. The diagram on the left below shows an 18 by 62 rectangular array. To the right of the array, 2 has been added to the width. At the bottom of the array, 2 has been deducted from the height. More area was gained by adding 2 to the width than was lost by reducing the height, so the estimate is too high.

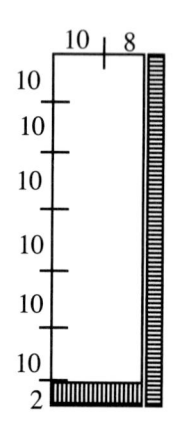

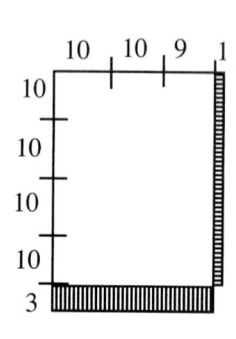

  b. An estimate for 43 × 29 would be 40 × 30 = 1200. This diagram shows a 43 by 29 array to represent the exact product. To the right of the array 1 has been added to the width. At the bottom of the array, 3 has been deducted from the height. More area was lost by deducting 3 from the height than was gained by increasing the width, so the estimate is too low.

33. a. Order of operations conventions tell us to multiply before adding, so we first find the products 62 × 45 and 14 × 29, then add the answers. To get an estimate we can take 60 × 50 and 10 × 30 (or 15 × 30) to get 3000 + 300 = 3300. The exact answer is 3196.

  b. In the computation of 36 + 18 × 40 + 15, we must first multiply 18 × 40, then do the addition. To estimate we could use 20 × 40 = 800, then add 35 + 800 + 15 = 850. The exact answer is 36 + 720 + 15 = 771.

35. a. The calculator will generate a geometric sequence because it is asked to multiply each previous answer by 3. The sequence is 5, 15, 45, 135, 405, 1215, 3645, 10935, 32805.

  b. Here we have an arithmetic sequence because we are continually adding 5. The sequence is 20, 25, 30, 35, 40, 45, 50, 55, 60, 65.

37. The key strokes are written here with view screen answers in **bold**.
  a. 84 × 52 = **42328**
  b. 1061 × 52 = **55172**
  c. 948 × 52 = **49296**
  d. 1198 × 52 = **62296**
  e. 996 × 52 = **51792**
  f. 1125 × 52 = **58500**

39. a. One first guess might be 30. 32 × 30 = 960. This is too large. Try 25. 32 × 25 = 800. Just barely in the range in two tries. *[Would 26 have been too large? 27?]*

b. Try 20. 95 × 20 = 1900. Too large. Try 15. 95 × 15 = 1425. Too small. Try 18. 95 × 18 = 1710. In the range in three tries.

41. a. The square numbers can be seen going down the diagonal from upper left to lower right. For every row there is a matching column -- for example, the second row is the same as the second column. *[What property does this relate to?]* There are also some matching diagonals. The units digits of the multiples of 6 are in the sequence 6, 2, 8, 4, 0, 6, 2, 8, 4, 0. Many other patterns can be found.

b. The units digits of the multiples of 9 count backward: 9,8,7,6,5,4,3,2,1. The sum of the digits is always 9.

43. If Harry pays for the car over time it will cost him the $500 down payment plus 24 monthly payments of $155 each. This is a total of 500 + (24 × 155) = $4220. But he will earn back $150 of this money with his investment. So his net cost this way is $4220 – $150 = $4070. This is quite a bit more expensive than just paying cash. He loses $4070 – $2500 = $1570.

45. Vanessa is only interested in the two largest sizes, call them large and jumbo. First we will just look at the large backpacks. There are 5 styles to choose from. For three of the five styles there are two choices of kind of material. Call the styles A,B,C,D, and E. Suppose styles A,B, and C have two kinds of material. Then Vanessa can choose from A1, A2, B1, B2, C1, C2, D, and E. So there are 8 large backpacks to choose from. There are also 8 jumbo backpacks to choose from. So Vanessa has 16 backpacks to choose from.
*[This problem is also a good one for drawing a sketch.]*

47. Here is the pattern:
$$1 \times 9 + 2 = 11$$
$$12 \times 9 + 3 = 111$$
$$123 \times 9 + 4 = 1111$$
$$1234 \times 9 + 5 = 11111$$
$$12345 \times 9 + 6 = 111111$$
The pattern continues to hold for the first 9 equations. After line 9 we have
12345678910 × 9 + 10 = 111,111,110,200 and the pattern is broken.
*[Why does this pattern work? Why does it fail after the 9th one?]*

64  Chapter 3  Whole Numbers

49. a. There are several patterns that one might notice in multiplying a two digit number by 99. Here are some examples:  54 × 99 = 5346;   61 × 99 = 6039;   25 × 99 = 2475.
1.) One pattern seems to be that the first two digits of the answer form a number that is one less than the number being multiplied.  2.) The last two digits are the difference between the original number and 100.  3.) The sum of the two-digit number formed by the first two digits and the two-digit number formed by the last two digits is 99.  Any of these three statements is a reasonable conjecture.  Others are also possible.
Here are some products of two-digit numbers by 999:       62 × 999 = 61938;   35 × 999 = 34965;     99 × 999 = 98901;    13 × 999 = 12987
From these products we might observe that: 4.) the first two digits of the product again form a number which is one less than the original number; 5.) the middle number is always 9;  6.) the last two digits are the difference between the original number and 100.

b. Testing the first conjecture above, we see that in 73 × 99 = 7227 the first two digits are 72, which is one less than 73.  And, the second conjecture holds too, because 27 is the difference between 100 and 73.  Conjectures #4,5, and 6 are valid in the example 46 × 999 = 45954.  *[Note: We have not yet proved any of the conjectures.  But the evidence seems to support them.  Any ideas why they work?]*

c. These conjectures need to be modified when we look at products of three-digit numbers by 99 and 999.  There are some patterns here, but they may be more difficult to spot.

51. Samir remembers the first two numbers, but he doesn't remember which is first and which is second.  If his luck is bad on choosing the right numbers then he'll first check it the wrong way.  Suppose he thinks the first two numbers are 17, 9, in that order, but they are really 9, 17.  He will try 17, 9, 1, then 17, 9, 2, etc.  Since the numbers go from 1 to 25 this will be 25 tries.  Then he will try all 25 ways starting with 9, 17.  In the worst case, the last try will work.  So the greatest number of different combinations that must be tried is 50.

53. For each boy chosen there will be 9 boy-girl combinations possible, because there are 9 girls available.  If they want between 70 and 80 combinations they need to choose 8 boys.  This would give 9 × 8 = 72 possible combinations.  More than 8 boys would give more than 80 possible combinations, but more could be chosen if not all possible combinations are used.

55. The left hand shows 7 and the right hand shows 8.  In the left hand, two fingers are raised and in the right hand 3 fingers are raised.  Since 2 + 3 = 5, the tens digit of 7 × 8 is 5.  The ones digit is found by multiplying the closed fingers.  This is 3 × 2 = 6.  So the answer is 56.

57. Set up the lattice multiplication as in the first diagram below.  In the second diagram, the products of 3 × 7, 4 × 7, 3 × 8, and 4 × 8 are filled in the four boxes with the digits separated.  In the third diagram we find the four digits of the answer to 34 × 78 by adding down the diagonals.  Start with the 2 in the lower right hand corner.  It is the units digit of the answer.  Then add 8 + 3 + 4 to get 15.  Write down the 5 and carry the 1 to the next diagonal.  Adding 1 + 2 + 1 + 2 gives 6.  The thousands digit is 2.  The answer is 2652.

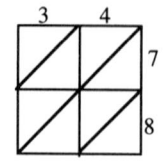

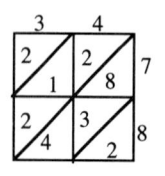

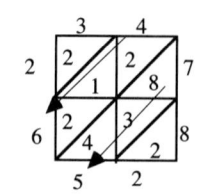

## Section 3.4

1. a. Using the sharing (or partitive) concept of division to illustrate 28 ÷ 7, we share the 28 dots in the diagram among 7 groups. Or, we partition them into 7 sets.

   b. Using the measurement (or subtractive) concept of division to illustrate 28 ÷ 7, we measure 7 dots into each group. This results in 4 groups.

3. A division exercise can be written as multiplication, because if a ÷ b = c, then c × b = a.
   a. 68 ÷ 17 = 4 can be written as 4 × 17 = 68.

   b. 414 ÷ 23 = 18 can be written as 18 × 23 = 414 (or, 23 × 18 = 414).

5. If we divide the product by either of the factors we will get the other factor. In symbols, if A × B = C, then C ÷ B = A (and, C ÷ A = B).
   a. We can write 14 × 24 = 336 as 336 ÷ 24 = 14. Or we can write 336 ÷ 14 = 24.

   b. We can write 9 × 8 = 72 as 72 ÷ 8 = 9. Or we can write 72 ÷ 9 = 8.

7. a. Using the sharing concept, dividing 396 by 3 means sharing the 396 units evenly among 3 groups. In this case we do not need to regroup. The 3 flats are shared one to each group, the 9 longs are divided with 3 going to each group, and the 6 units are shared with two going to each group. See the diagram on the left below.

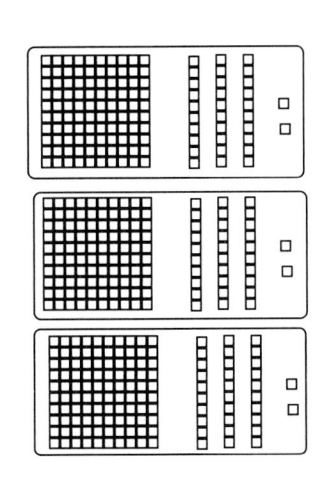

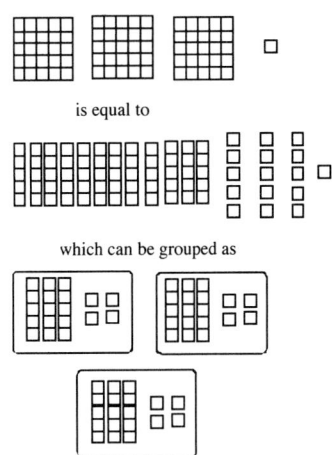

   b. The figure above on the right shows $301_{five}$. It is regrouped as 12 longs and 16 units. These are then shared among three groups. The answer is $301_{five} ÷ 3 = 34_{five}$.

9. a. We can not divide 3 flats into 7 groups (or into groups of 7), so we trade the 3 flats for 30 longs. The 30 longs together with the original 9 longs make 39 longs which can be divided into 7 groups of 5, leaving 4 extra longs. Trade these four longs for units and we have 40 + 2 = 42 units. These can be divided into 7 groups of 6. In the end we have 7 groups of 56. This is shown in the diagram below.

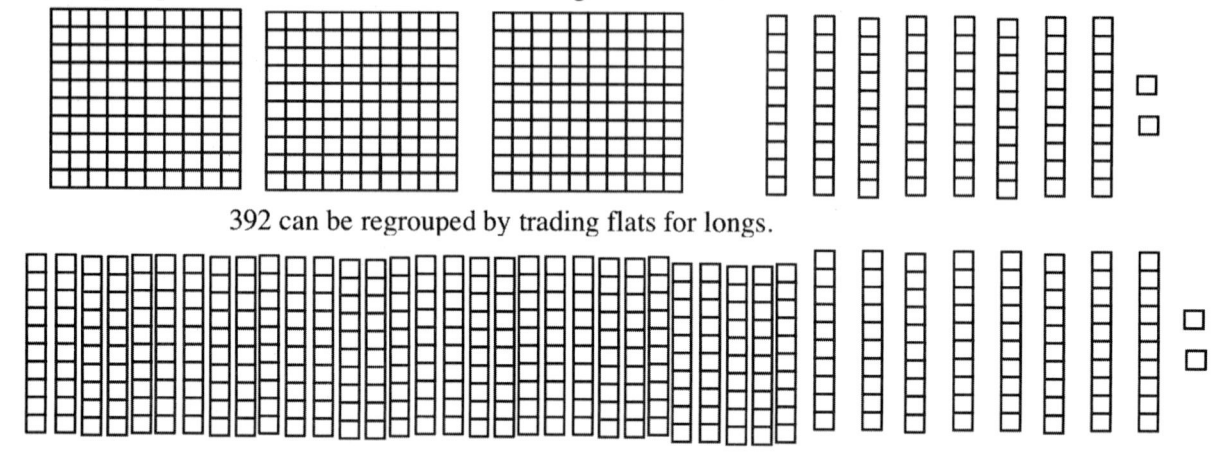

392 can be regrouped by trading flats for longs.

Then 35 of the 39 longs can be divided into 7 groups of 5 with the other 4 longs traded for units.

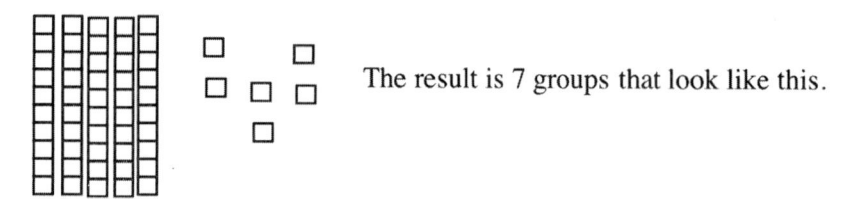

The result is 7 groups that look like this.

b. The diagram for 320 ÷ 5 is similar. Start with 3 flats and 2 longs. Trade the 3 flats for 30 longs. Then we have 32 longs. Put 30 of them into 5 groups of 6. There are 2 longs left. Trade these for 20 units. Put them in 5 groups of 4. Together there are 5 groups, each containing 6 longs and 4 units. So 320 ÷ 5 = 64.

11. The diagrams below show the rectangular arrays. The array for 72 ÷ 12 has an area of 72 and one dimension of 12. The other dimension is the quotient 6. For 286 ÷ 26 the rectangle has area 286 and dimensions 26 by 11.

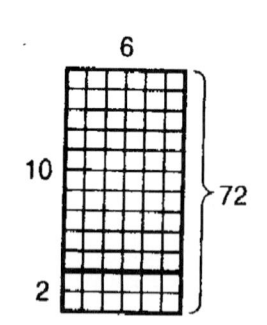

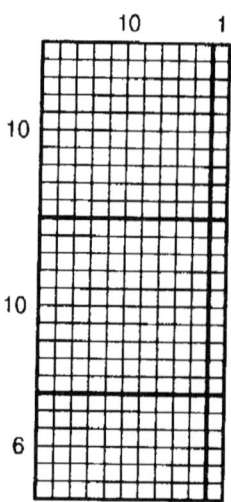

*3.4 Division and Exponents* 67

13. a. We need a rectangle with area 608 and one dimension 32. We will need to regroup. With only 6 flats and 8 units we can not make a rectangle with one dimension 32. Place 3 of the flats in a row to make a 10 by 30 rectangle. Trade the other 3 flats for 30 longs. Two of the longs are needed to make a 10 by 32 rectangle. Of the remaining 28 flats, 27 can be used to extend the rectangle to 32 by 19, if in addition the remaining long is traded for 10 units, so that the 18 units fill the remaining space. See the completed rectangle below.

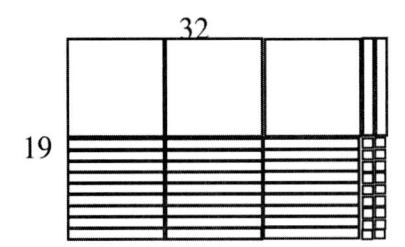

$19 \times 32 = 608$

$608 \div 32 = 19$

b. To create a rectangle with area 221 and one dimension 13, we will again need to trade for some more longs. Trade one flat for 10 longs and 2 longs for 20 units. Now there are 1 flat, 10 longs, and 21 units. For a dimension of 13, place 3 of the longs below the flat. Then place the remaining 7 longs to the right of the flat. That leaves room for the 21 units to complete a 13 × 17 rectangle. (*See below at left.*)

#13b
$13 \times 17 = 221$
$221 \div 13 = 17$

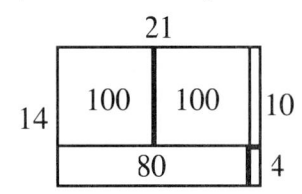

#13c
$14 \times 21 = 294$
$294 \div 21 = 14$

c. This time, for 294 ÷ 21 we do not need more longs. Place the two flats next to each other with one long next to them. This gives a 10 × 21 rectangle. Place the remaining 8 longs in two groups of four, along with the 4 units to complete a 14 × 21 rectangle. (*See above.*)

15. a. 0 ÷ 4 can be computed. It is equal to 0.

    b. 4 ÷ 0 can not be computed. There is no number that can be multiplied by 0 to give 4.

    c. 0 ÷ 0 can not be computed.

17. a. Since 5 spaces are subtracted repeatedly from 15, this diagram shows 15 ÷ 5.

    b. The diagram below shows 18 ÷ 6 because 6 spaces are subtracted repeatedly from 18.

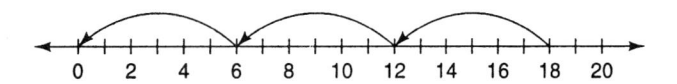

19. a. In dividing 4052 by 8, there are 40 hundreds which gives 5 hundreds when divided by 8. In this example the 5 was placed in the hundreds place, but then by neglecting to put a zero in the tens place, it looks like the answer is 56 R4. But it should be 506 R4.

    b. The answer should be 86, not 68. The 8 and 6 were placed in the wrong place values.

21. a. This equation is true. For example, $(9 + 6) \div 3 = 9 \div 3 + 6 \div 3$, because $15 \div 3 = 3 + 2$. There are no counterexamples for this one because division is distributive over addition.

    b. Here we can find a counterexample such as $8 \div 4 = 2$ but $4 \div 8 = .5$. In fact any time the two numbers are different the two sides will not be equal. Division is not commutative.

23. a. Addition is not closed on the set of odd whole numbers. For example, 5 and 3 are both odd numbers, but $5 + 3 = 8$ and 8 is not in the set of odd whole numbers. In fact the sum of two odd numbers is always even. [*Could you prove this with a picture?*]

    b. Division is not closed on the set of whole numbers. For example, 4 and 7 are both whole numbers, but the quotient $4 \div 7$ is not a whole number.

    c. Multiplication is closed on the set of whole numbers. The product of any two whole numbers is always another element of the set of whole numbers.

25. a. On the calculator, $47208 \div 674$ gives 70.04154 etc. Multiplying $70 \times 674$ gives 47180. The difference between 47208 and 47180 is 28. So the quotient is 70 R28.

    b. Dividing $2018 \div 17$ on the calculator gives 118 and some decimal. Multiplying $118 \times 17$ gives 2006, so the remainder is 12.

    c. Dividing $1121496 \div 465$ on the calculator gives 2411 and some decimal. Multiplying $2411 \times 465$ gives 1121115, so the remainder is $1121496 - 1121115 = 381$.

27. a. Both 90 and 18 are multiples of 9, so the quotient of $90 \div 18$ will be the same as the quotient of $10 \div 2$. The answer is 5.

    b. Both 84 and 14 are even numbers. We can divide both by two and have an equal quotient. $84 \div 14 = 42 \div 7 = 6$.

29. a. Rounding 46 to 50 we get $250 \div 46 \approx 250 \div 50 = 5$. This answer will be less than the exact quotient because we rounded the divisor up.

    b. Rounding to the nearest 10, we get $82 \div 19 \approx 80 \div 20 = 4$. The divisor was rounded up and the dividend was rounded down. The estimate is less than the actual quotient.

    c. This time we will round the divisor down and the dividend up. $486 \div 53 \approx 500 \div 50 = 10$. The estimate will be greater than the actual quotient.

31. a. Using front-end estimation we replace all but the leading digits with zeros.
    $623 \div 209 \approx 600 \div 200 = 3$.

    b. $7218 \div 1035 \approx 7000 \div 1000 = 7$.

33. a. When the bases are the same we can multiply by adding the exponents. $5^{14} \times 5^{20} = 5^{34}$.

    b. $10^{12} \div 10^{10} = 10^2$.

35. a. We need to multiply before adding and subtracting left to right.
    $6 + 4 \times 8 - 3 = 6 + 32 - 3 = 38 - 3 = 35$.

    b. First multiply $5 \times 10$ and $2 \times 6$. This gives $50 - 12 = 38$.

    c. First perform the operation in parentheses. Then multiply left to right.
    $5 \times (10 - 2) \times 6 = 5 \times 8 \times 6 = 40 \times 6 = 240$.

    d. Divide and multiply left to right. Subtract last.
    $45 \div 3 \times 5 - 2 = 15 \times 5 - 2 = 75 - 2 = 73$.

37. a. To find how many times greater, we need to divide $10^{21}$ by $10^8$. Since the bases are the same, this can be accomplished by subtracting the exponents. $21 - 8 = 13$.
    The frequency of the gamma rays is $10^{13}$ times the radio frequency.

    b. Divide $10^{20}$ by $10^5$ to see that the gamma ray frequency is $10^{15}$ times the frequency of a long wave.

39. This sequence of calculator steps will produce the correct answer because $8 \times (12 \div 3)$ is equal to $(8 \times 12) \div 3$.

41. On some calculators this sequence will produce the correct answer. On some it will not. It depends whether the order of operations has been programmed into the calculator.

43. a. 1. Q, 4 and R, 4
       2. Q, 3 and R, 5
       3. Q, 4 and R, 6
       4. Q, 4 and R, 2
       5. Q, 3 and R, 6
       6. Q, 3 and R, 3

    b. If there is no mixing of classes, a total of $5 + 4 + 5 + 5 + 4 + 4 = 27$ vans are needed.

45. a. Each answer is found by dividing the previous one by seven. The six results are:
       1. 1,647,086
       2. 235,298
       3. 33,614
       4. 4802
       5. 686
       6. 98

    b. The ninth number will be less than one.

47. a. Q, 510 and R, 13.      b. Q, 12 and R, 406.

49. a. The calculator will first find 10 to the sixth power, which is 1,000,000. Then it will divide by two, and finally add 14. The result is 500014.

   b. The calculator first finds 9 to the $4^{th}$ power, then adds 251. The result is 6812.

51. The base numbers on the left sides of the equations are in the pattern:
    $1 \times 2 = 2$; $2 \times 3 = 6$; $3 \times 4 = 12$. So the next one should be $4 \times 5 = 20$. The base number on the right side of each equation is the next consecutive number after the third base on the left. Each number is raised to the second power.
    The $4^{th}$ equation is: $4^2 + 5^2 + 20^2 = 21^2$
    The $12^{th}$ equation is: $12^2 + 13^2 + 156^2 = 157^2$

53. There are a number of patterns that can be found in this triangle of numbers. Here are some:
    The $n^{th}$ row is a sum of n consecutive odd numbers.
    The sum of the $n^{th}$ row is the cube of the number n.
    The middle number in the $3^{rd}$ row is 3 squared. The middle number in the $5^{th}$ row is $5^2$.
    The middle number in the $11^{th}$ row will be $11^2 = 121$. The $11^{th}$ row will contain 11 numbers.
    Here is the $11^{th}$ row:
    $111 + 113 + 115 + 117 + 119 + 121 + 123 + 125 + 127 + 129 + 131 = 1331$
    Here is the $12^{th}$ row:
    $133 + 135 + 137 + 139 + 141 + 143 + 145 + 147 + 149 + 151 + 153 + 155 = 1728$

55. a. The units digits in powers of 2 follow this pattern: 2, 4, 8, 6, 2, 4, 8, 6, . . .
    The $100^{th}$ power will end in 6, because every $4^{th}$ one ends in 6. So the $101^{st}$ will end in 2, the $102^{nd}$ in 4, and the $103^{rd}$ in 8.

   b. The units digit of a power of 6 is always 6, because $6 \times 6 = 36$. Every time another 6 is multiplied the units digit is always found from the units digit in $6 \times 6$.

57. a. To decide which is the greater payment we need to add $1 + $2 + $4 etc. for 22 weeks. It might help to look at the first few weeks and see if there is a pattern.
    $1 + 2 = 3$; $1 + 2 + 4 = 7$; $1 + 2 + 4 + 8 = 15$; $1 + 2 + 4 + 8 + 16 = 31$.
    Note that each of these sums is one less than a power of two. In fact it is always one dollar less than the next weeks pay. The total after two weeks is $2^2 - 1$, after three weeks it is $2^3 - 1$, etc. After 22 weeks it will be $2^{22} - 1 = \$4{,}194{,}303$. So the doubling each week method results in a greater payment.

   b. The total amount from doubling each week is $2,194,303 greater than $2,000,000.

59. If each tree gives a gallon of sap every five days, then in 30 days each tree will give 6 gallons of sap. There are 15 trees, so over the 30 day period they should produce $15 \times 6 = 90$ gallons of sap. The 90 gallons when boiled down produce 1 gallon of syrup for every 40 gallons, so there will be $90 \div 40 = 2.25$ gallons of syrup. There are 4 quarts in a gallon, so this is exactly 9 quarts. Divided among 3 people they should each get 3 quarts.

## Chapter 3 Test

1. a. The Egyptian notation is additive. In 226 there are 2 100's, 2 10's, and 6 1's.

    ϙϙ∩∩||||||

   b. For Roman numeration, think of 226 as 200 + 20 + 5 + 1. It is CCXXVI.

   c. Babylonian numeration is based on sixties. 226 has three 60's and 46 ones.

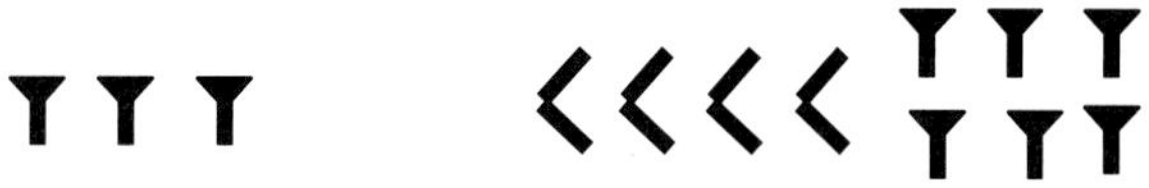

     ᛉ ᛉ ᛉ         ⟨⟨⟨⟨ ᛉ ᛉ ᛉ
                         ᛉ ᛉ ᛉ

   d. The Mayan numeration is based on twenties. 226 has 11 twenties and 6 ones.

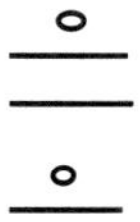

2. a. The 4 is underlined and is in the millions place. Its value is 4,000,000.

   b. A zero is underlined in the ten-thousands place. It has value 0 because 0 × 10,000 = 0.

3. a. The number 6,281,497 is closer to 6,300,000 than 6,200,000.

   b. The number 6,281,497 is closer to 6,281,500 than 6,281,400.

   c. The number 6,281,497 is closer to 6,281,000 than 6,282,000.

4. a.

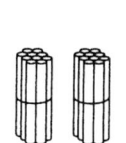

   b.

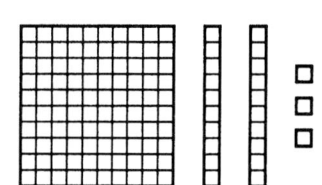

   c.

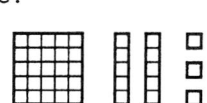

   d.

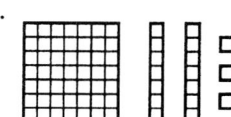

5. a. To add 245 + 182 using base 10 pieces we need to combine pieces and then regroup. In the sum there are 12 longs, so 10 of them can be traded for a flat. The sum then contains 4 flats, 2 longs, and 7 units. The diagram below shows that 245 + 182 = 427.

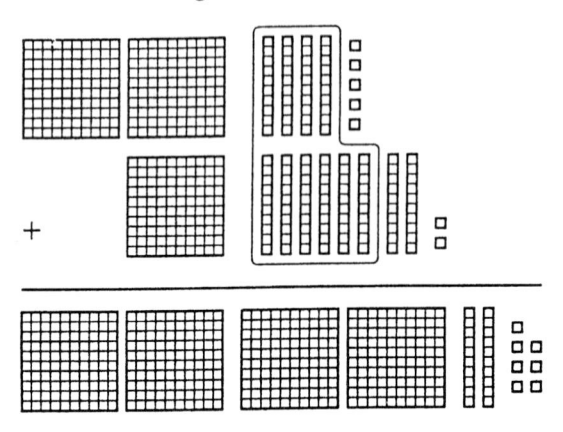

   b. To subtract 362 – 148 using base 10 pieces we need to trade a long for 10 units. Using the take-away model we can take 8 units from 12 units, take 4 longs from 5 longs, and take 1 flat from 3 flats. The result is 2 flats, 1 long, and 4 units. So 362 – 148 = 214.

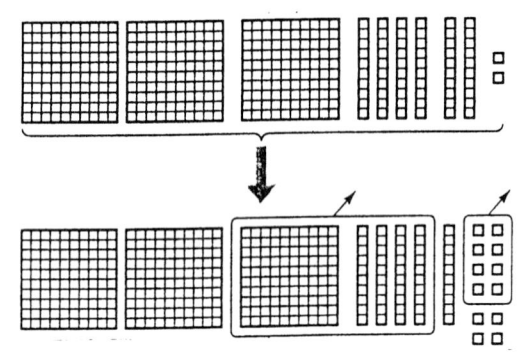

6. a. Using left-to-right addition, we first add the hundreds, then the tens, and last the units. There are 15 tens, so we must add one more to the hundreds and leave 5 tens.
   483 + 274 = 600 + 150 + 7 = 757.

   b. When adding with partial sums we write each place value on a separate line, then we add.
   ```
        864
      + 759
         13
         11
         15
       1623
   ```

7. a. To use equal differences we add or subtract the same amount to both numbers to get another equivalent expression for the difference that is easier to compute mentally. One way here is to add 1 to both numbers. 65 – 19 = 66 – 20 = 16.

   b. Adding three to both numbers gives 843 – 97 = 846 – 100 = 746.

8. In using front-end estimation with the leading digits we replace all digits except the first one with zeros and then estimate by calculating with these numbers.

   a. 321 + 435 + 106 ≈ 300 + 400 + 100 = 800.

   b. 7410 – 2563 + 4602 ≈ 7000 – 2000 + 4000 = 9000.

   c. 32 × 56 ≈ 30 × 50 = 1500.

   d. 3528 ÷ 713 ≈ 3000 ÷ 700 ≈ 4. (A closer approximation here would be 3500 ÷ 700 = 5.)

9. a. An equal product will <u>not</u> be obtained by multiplying both factors by the same number, but if we multiply one factor by a number and divide the other factor by the same number we will have an equal product. A convenient way on this one is: 18 × 5 = 9 × 10 = 90. Essentially we are using the associative property: 18 × 5 = (9 × 2) × 5 = 9 × (2 × 5).

   b. Multiply 25 by 4 and divide 28 by 4: 25 × 28 = 100 × 7 = 700.

10. The partial products in 43 × 28 consist of 3 × 8 = 24, 3 × 20 = 60, 40 × 8 = 320, and 40 × 20 = 800.

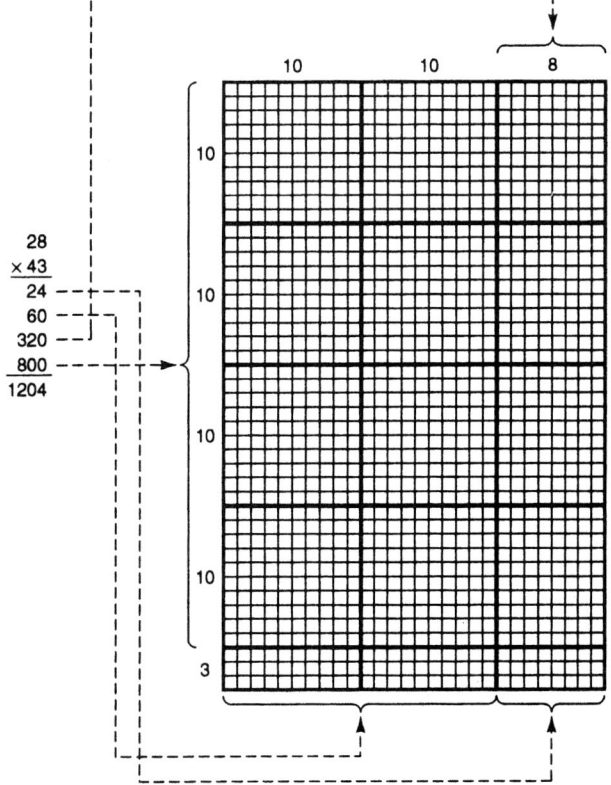

11. a. We multiply first, left to right, and then subtract.
    $6 \times 4 \times 5 - 3 = 24 \times 5 - 3 = 120 - 3 = 117$.

    b. First multiply and divide. Then add the two results.
    $48 \div 4 + 2 \times 10 = 12 + 20 = 32$.

    c. First perform the operation in parentheses, then multiply, and subtract last.
    $(8 + 3) \times 5 - 2 = 11 \times 5 - 2 = 55 - 2 = 53$.

12. a. If base numbers are the same we perform the division by subtracting the exponents.
    $3^{12} \div 3^4 = 3^8$.

    b. The base numbers are equal so we perform the multiplication by adding the exponents.
    $7^4 \times 7^6 = 7^{10}$.

13. We show the sharing concept for $452 \div 4$ by dividing the 4 flats, 5 longs, and 2 units into 4 equal groups. To do this we need to trade one of the longs for 12 units. The diagram below shows the division with base 10 pieces.

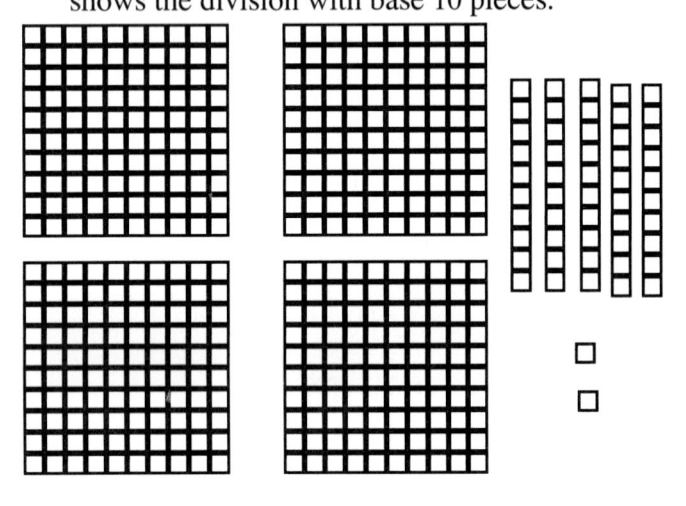

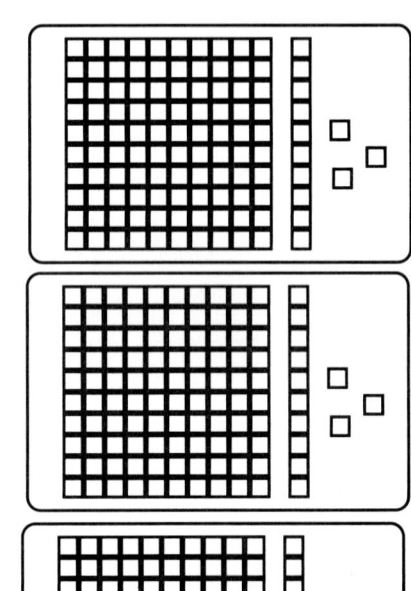

14. There are several ways to mentally estimate each of these computations.

   a. Rounding to hundreds gives: $473 + 192 \approx 500 + 200 = 700$.
      Or, a more exact rounding method is: $470 + 200 = 670$.
      Or, an exact answer using compatible numbers is: $473 + 192 = (473 - 8) + (8 + 192)$
      $$= 465 + 200 = 665.$$

   b. Rounding to hundreds gives: $534 - 203 \approx 500 - 200 = 300$.
      Or, $534 - 203 \approx 530 - 200 = 330$.

   c. Since 993 is close to 1000, a reasonable estimate is $993 \times 42 \approx 1000 \times 42 = 42{,}000$.
      Another reasonable estimate is $1000 \times 40 = 40{,}000$.

   d. We can round 49 to 50 to get: $350 \div 49 \approx 350 \div 50 = 7$.
      Or, using equal quotients, $350 \div 49 = 50 \div 7 \approx 7$.

15. a. The statement is true. Multiplication is distributive over subtraction. For example,
      $7 \times (5 - 3) = 7 \times 5 - 7 \times 3$ because $7 \times 2 = 14$ and $35 - 21 = 14$.
    b. It is true that addition is commutative. For example, $8 + 5 = 5 + 8$.
    c. Subtraction is not associative. Here is a counterexample: $(10 - 6) - 2 \neq 10 - (6 - 2)$.
    d. Division is not commutative. Here is a counterexample: $10 \div 2 = 5$ but $2 \div 10 = 0.2$.
    e. This statement is false. Subtraction is not closed on the whole numbers because we can find pairs of whole numbers that do not give whole numbers when subtracted. For example, $5 - 7 = -2$ is not in the set of whole numbers.

16. Here is the pattern of equations with the fourth equation included.
    $$3^2 + 4^2 = 5^2$$
    $$10^2 + 11^2 + 12^2 = 13^2 + 14^2$$
    $$21^2 + 22^2 + 23^2 + 24^2 = 25^2 + 26^2 + 27^2$$
    $$36^2 + 37^2 + 38^2 + 39^2 + 40^2 = 41^2 + 42^2 + 43^2 + 44^2$$
    There are various ways to discover this pattern and various things to notice in the pattern. The number of terms on each side of the equation grows by one each time. The first base number increases by 7, then 11, then 15, suggesting a use of finite differences to continue the pattern. Finite differences can also be used in other parts of this pattern. The pattern does hold for the fourth equation. Both sides are equal to 7230. *[Does it continue?]*

17. There were 49 football players and 18 baseball players, for a total of $49 + 18 = 67$ players of these two sports. Since there were 61 total athletes who played either football or baseball or both, there must have been $67 - 61 = 6$ athletes who played both. The Venn diagram below shows the situation.

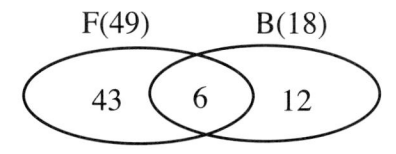

    F(49)   B(18)
    
    43   6   12

18. There are 4 different sizes of pizzas and (counting plain) 6 different ways to top them. So there are $4 \times 6 = 24$ different types of pizza. *[What if combinations are allowed?]*

# Chapter 4 Number Theory
## Section 4.1

1. If you wanted to make your one flight walked going down, then the solutions below will work. There are other solutions that are just as good. If you wanted a little exercise other solutions may be better.

   a. Take the escalator to the elevator that serves the even-numbered floors and deliver to the 26$^{th}$ and 48$^{th}$ floors. Then walk down to the 47$^{th}$ floor and use the elevator that serves the odd-numbered floors to deliver to the 35$^{th}$ and 11$^{th}$ floors. Return to the street level on the elevator that serves the odd-numbered floors.

   b. Use the top deck elevator to deliver to floors 20, 22, 24, 26, 28, and 30. Then walk down one flight, deliver to floor 29, and use the bottom deck elevator to deliver to 27, 25, 23, and 21.

3. a. The statement 3 | 4263 is true because 3 divides 4263. We can check that 3 is a factor of 4263 by adding the digits 4 + 2 + 6 + 3 = 15. Since 3 divides 15, 3 | 4263.

   b. The statement 15 | 1670 is false. 15 does divide 1500 and 15 also divides 150, so we know that 15 divides the sum 1500 + 150. That is, 15 | 1650. Then it will be true that 15 | 1665 and false that 15 | 1670.

   c. The statement here reads "12 does not divide 84". This statement is false because 12 × 7 = 84 so 12 is a factor of 84.

5. a. 7 | 63    b. 8 | 40    c. 13 | 39    d. 12 | 36

7. a. A linear model to show that 6 divides 54 shows a line segment that is 54 units long which is subdivided evenly into 6-unit long segments.

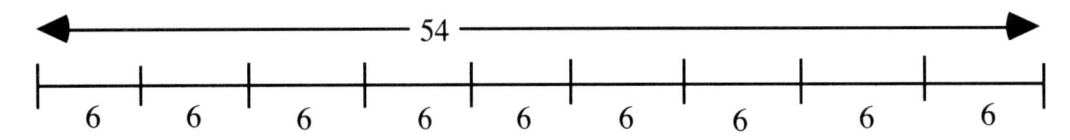

   ```
   ◄─────────── 54 ───────────►
   ├───┼───┼───┼───┼───┼───┼───┼───┼───┤
     6   6   6   6   6   6   6   6   6
   ```

   b. An area model to show divisibility shows the area divided by one of the dimensions. To show that 12 | 60 we can show a rectangle with area 60 and one dimension 12. If the other dimension is a whole number, then it is divisible. In this case the other dimension is 5.

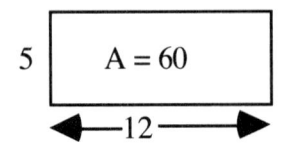

   ```
       ┌──────────┐
     5 │  A = 60  │
       └──────────┘
        ◄── 12 ──►
   ```

## 4.1 Factors and Multiples

9. a. Rods whose numbers are factors (divisors) of 12 will form trains for 12. Besides using 6 red rods, we can also use 12 white rods or 4 green rods or 3 purple rods or 2 dark green rods.

   b. One-color trains to represent 15 could be made from white rods, green rods, or yellow rods. This shows that 15 is evenly divisible by 1, 3, or 5. (It is also divisible by 15, but there are no rods that are 15 units long.)

   c. For each of the numbers 2, 3, 5, and 7 there will be two different one-color trains. There will be a white train and a train made up of just one rod of that number's color. This illustrates the fact that any prime number has exactly two factors, one and itself.

11. a. Prime numbers will have only one rectangular array possible. Its dimensions will be 1 by p for any prime number p.

    b. The numbers 15, 30, and 17 have an even number of factors. The factors of 15 are 1, 3, 5, and 15. The factors of 30 are 1, 2, 3, 5, 6, 10, 15, and 30. The factors of 17 are 1 and 17. The rectangular arrays for 15, 30, and 17 are shown below.

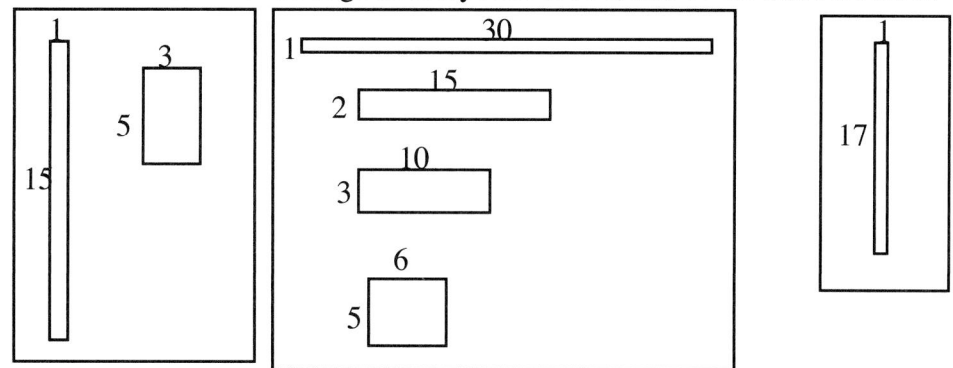

    c. If a number does not have a square array, then all of its arrays are non-square rectangles which have two different factors. We can think of each set of two factors as a pair.

    d. If a number has 8 factors then it has 4 different rectangular arrays. As we see in part b above, 30 is one example of a number with 8 factors. There is a smaller number with exactly 8 factors. Notice that the shorter dimensions of the rectangles for 30 are 1,2,3,5. What if we look at four rectangles with smallest dimensions 1, 2, 3, and 4. Is it possible to use all of these as dimensions for different rectangles that all have the same area **and** this area is less than 30? Twenty does not work because 3 is not a factor of 20. How about 24? Rectangles with area 24 would be 1 × 24, 2 × 12, 3 × 8, and 4 × 6. All whole numbers less than 24 have less than 8 factors.

13. a. The sum of the digits of 465,076,800 is 4 + 6 + 5 + 7 + 6 + 8 = 36. Since 3 | 36, 3 is also a divisor of 465,076,800. When 465,076,800 is divided by 3 the remainder will be 0.

    b. The number 100,101,000 is also divisible by three because the sum of the digits is 3.
    *[Why does this divisibility rule work?]*

15. a. The sum of the digits in 48,276,348,114 is 48. Since 9 is not a divisor of 48 it is also not a divisor of 48,276,348,114. Since the remainder is 3 when dividing 48 by 9, the remainder is also 3 when dividing 48,276,348,114 by 9.

b. The sum of the digits in 206,347,166,489 is 56. Since 9 is not a divisor of 56 it is also not a divisor of 206,347,166,489. Since the remainder is 2 when dividing 56 by 9, the remainder is also 2 when dividing 206,347,166,489 by 9.

17. a. The number 12 is divisible by 3, but not by 9. This provides a counterexample showing that the answer to "If a number is divisible by 3, is it divisible by 9?" is no.

b. On the other hand, it is true that if a number is divisible by 9 then it is divisible by 3. Since 9 itself is a multiple of 3, any multiple of 9 such as 18, 27, etc. is also a multiple of 3. For example, 18 = 9 × 2 = 3 × 3 × 2 = 3 × 6.

19. a. To test for divisibility by 4 we need to look at the last two digits of the number. This works because 4 | 100 and so 4 divides any multiple of 100. For example, 4 | 536 because 4 | 500 and 4 | 36. To test whether 4 | 47,382,729,162 we just need to check whether or not 4 | 62. Since 4 × 15 = 60, 4 does not divide 62. The remainder in 62 ÷ 4 is 2. So 4 is not a divisor of 47,382,729,162, and when 47,382,729,162 is divided by 4 the remainder is 2.

b. Since 4 | 76 we also know that 4 | 512,112,911,576.

21. In the diagram on the left below, note that a base 10 flat is evenly divisible by 4, but neither a long nor a unit are divisible by 4. Any base 10 piece larger than a flat, such as a long-flat will also be divisible by 4, because it is made up of flats. So to determine divisibility by 4 we just need to look at the longs and units. For example, look at the base 10 pieces for the number 1,324. They are shown (on a smaller scale) in the diagram on the right. We know that 4 | 1300 because 4 | 100. Since it is also true that 4 | 24, we also know that 4 | 1324.

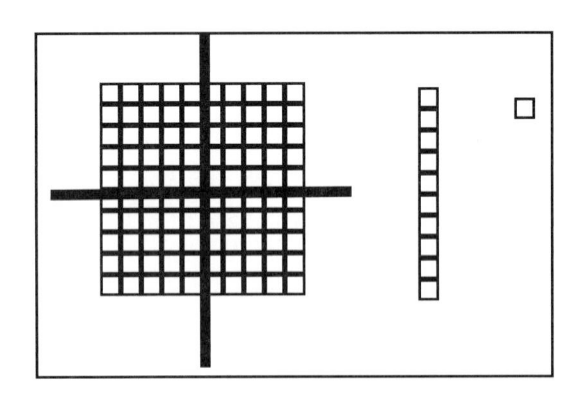

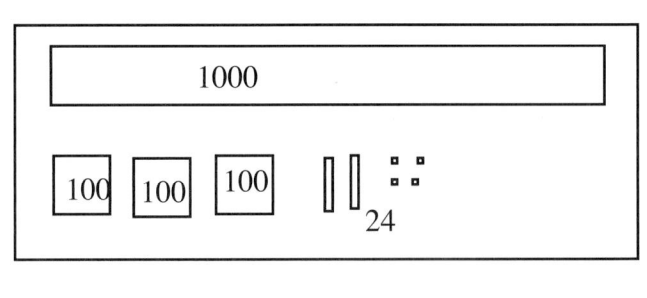

b. In words the statement is: "If a does not divide b and a does not divide c, then a does not divide the sum of a and b". This statement is false. We can find a counterexample. For one counterexample let a = 5, b = 7, and c = 8. Neither 7 nor 8 are divisible by 5, but their sum is 15, which is divisible by 5. Many other counterexamples can be found.

23. a. In words the statement is: "If a divides b and a does not divide c, then a does not divide the difference between b and c". This statement is true. The diagram below shows why.

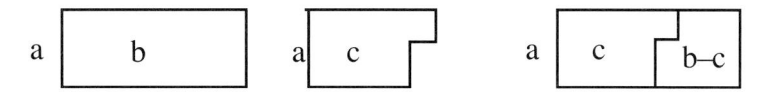

c. In words the statement is: "If a divides b and b divides c, then a divides c". Another way to say this is: "If b is a multiple of a and c is a multiple of b, then c is a multiple of a". These statements are true. The diagram below shows the situation when b = 4a and c = 3b. In that case, c = 12a.

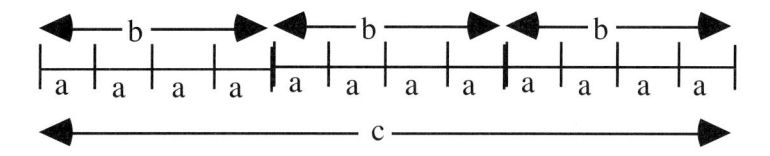

25. If none of 2, 3, 5, 7, 11, and 13 are factors of 173, then no multiples of these numbers are factors of 173 either. That means that no numbers between 2 and 13 are multiples of 173 because these are all the primes up to 13. The next prime number is 17. That would be the next number we would need to check to see if it is a factor of 173. But 17 × 17 = 289, so if 17 is a factor of 173 then the other number in the factor pair is less than 17. But we already know that there are no numbers smaller than 17 that are factors of 173. The last number we really needed to try was 13. So 173 is prime.

27. a. To check whether a number is prime we need to check only for prime number factors. The number 231 is not divisible by 2 because it is not even. However it is divisible by 3 because the sum of the digits is 6. So 231 is not prime.

b. Again we check for prime factors. The number 227 is not divisible by 2, 3, 5, 7, 11, or 13. This is as far as we need to check because the square of the next prime, 17, is 289. Since 227 has no prime factors less than 17, it will have no factors other than 1 and itself. So 227 is a prime number.

c. To test whether 683 is prime we will need to check prime numbers up to 23. The square of 23 is 529. The next prime number is 29, whose square is 841. So, if 683 has any factors (other than 1 and 683), it will need to be divisible by at least one of the following prime numbers: 2, 3, 5, 7, 11, 13, 17, 19, or 23. Through the use of divisibility tests and/or a calculator we see that none of these are divisors and 683 is prime.

29. To determine all of the primes less than 300 the Sieve of Eratosthenes needs to be extended by including all of the numbers up to 300. Then in addition to crossing out all multiples of 2, 3, 5, and 7, we would also cross out multiples of 11, 13, and 17. We would not need to look at multiples of 19 because the first one not already crossed out would be 19 × 19 = 361.

31. a. To use this test for divisibility by 11 we start in the units place and alternately add and subtract digits. To test whether 11 divides 63,011,454 we compute:
4 – 5 + 4 – 1 + 1 – 0 + 3 – 6. The result is 0.
Since 0 is divisible by 11 (0 ÷ 11 = 0), the number 63,011,454 is also divisible by 11.

   b. Applying the same test to 19,321,488 we get: 8 – 8 + 4 – 1 + 2 – 3 + 9 – 1 = 10.
Since 10 is not a multiple of 11, neither is 19,321,488.

   c. Applying the same test to 4,209,909,682 gives: 2 – 8 + 6 – 9 + 0 – 9 + 9 – 0 + 2 – 4 = –11.
It is also true that 11 divides –11 (11 × –1 = –11), so 11 | 4,209,909,682.

   d. The test will still work if we alternately add and subtract left to right. In doing this we may end up adding the numbers we subtracted going the other way, and subtract the numbers we added before. If this is the case then our result will be the opposite number. For example, if we get 7 going right to left, we might get –7 going left to right. Both a number and its opposite will tell us the same thing about divisibility by 11.

33. 17 + 2 = 19. 19 + 4 = 23. 23 + 6 = 29. 29 + 8 = 37. . . .
Continue adding the next even number to the previous result to get the next term in the sequence. Eventually we get to 199 + 28 = 227. 227 + 30 = 257. 257 + 32 = 289.
The sequence is 17,19,23,29,37,47,59,73,89,107,127,149,173,199,227,257,289.
All of these are prime until we get to 289. It is the square of 17.
*[We started with 17 and 289 is the 17$^{th}$ number in the list. Is this all coincidence?]*

35. When n = 1, the formula $n^2 – n + 41$ gives 1 – 1 + 41 = 41. When n = 2 we get
4 – 2 + 41 = 43. For n = 3 we get 9 – 3 + 41 = 47. For n = 4 we get 53. For n = 5 we get 61. For n = 6 we get 71. For n = 7 we get 83. For n = 8 we get 97. Larger values of n give us numbers over 100. *[Is there a connection here with the pattern in #33?]*

37. We need to check for at least two primes between 6 and 12, between 7 and 14, 8 and 16, etc. up to between 14 and 28. The primes 7 and 11 are between 6 and 12; 11 and 13 are between 7 and 14, 8 and 16, 9 and 18, and 10 and 20. Similarly, we can find at least two primes between the other pairs of numbers.

39. We want to compare the sum 33 + 35 + 37 + . . . + 499 + 501
with the sum 32 + 34 + 36 + . . . + 498 + 500.
Note that 33 is one more than 32, 35 is one more than 34, etc. all the way up to 501 which is one more than 500. How many numbers are there in each of these sums? There are 50 odd numbers from 1 to 99 inclusive, so there are 50 – 16 = 34 odd numbers from 33 to 99. There are 50 more odds from 101 to 199 inclusive, 50 from 201 to 299, 50 from 301 to 399, 50 from 401 to 499, and one more to include 501. This is a total of 235 odd numbers in the sum 33 + 35 + 37 + . . . + 499 + 501. There are also 235 even numbers in the sum 32 + 34 + 36 + . . . + 498 + 500, each of which is one less than its corresponding odd number from the sum of odds. So the sum of the evens is 235 less than the sum of the odds. Another approach to this problem would be to draw staircases to represent the sums and compare them.

## 4.1 Factors and Multiples

41. If we include any even numbers as factors, the product will be even because the product of two evens is even and the product of an even times an odd is even. So our only possible factors are 3, 5, 7, 11, and 13. The product of these numbers is 3 × 5 × 7 × 11 × 13 = 15015. Another approach to this problem is to note that 15015 passes the divisibility tests for 3, 5, and 11. Then some trial and error will find the other factors.

43. a. None of the numbers in the given sequence can be prime. The first one is divisible by 2 because both the product 2 × 3 × 4 × 5 × 6 and the number 2 are divisible by 2. So their sum must also be divisible by two. Similarly, 2 × 3 × 4 × 5 × 6 + 3 is divisible by 3, etc. To construct a sequence of 10 consecutive whole numbers that includes no primes we can use a similar pattern but make the product include the 10 consecutive numbers from 2 to 11. Then we can add the numbers from 2 through 11 to the product to produce our 10 consecutive composite numbers. Here is the sequence.

    2 × 3 × 4 × 5 × 6 × 7 × 8 × 9 × 10 × 11 + 2
    2 × 3 × 4 × 5 × 6 × 7 × 8 × 9 × 10 × 11 + 3
    2 × 3 × 4 × 5 × 6 × 7 × 8 × 9 × 10 × 11 + 4
    2 × 3 × 4 × 5 × 6 × 7 × 8 × 9 × 10 × 11 + 5
    •
    •
    •
    2 × 3 × 4 × 5 × 6 × 7 × 8 × 9 × 10 × 11 + 11

    b. To construct a sequence of 100 consecutive composite numbers we could use the same pattern. We would then use the product 2 × 3 × 4 × . . . × 100 × 101 and in turn add the whole numbers from 2 through 101. The first number would be divisible by 2, the second by 3, etc. all the way to the 100$^{th}$ number which would be divisible by 101.

45. The first two digits of the license plate number form a prime number, with the tens digit one less than the units digit. The numbers that satisfy these criteria are 23, 67, and 89. The sum of the digits of this number must also form a two-digit prime number. Only 67 satisfies that requirement. To find the three-digit number, let's first use the facts that the digits are three different odd numbers and their sum is palindromic. Possible sums for different combinations of three of 1, 3, 5, 7, and 9 are the odd numbers between 9 and 21, since 1+3+5 = 9 and 5+7+9 = 21. Of these, the only palindromic number is 11. So the three digits must be 1, 3, and 7. For the sum of the first and second digits to be twice the sum of the second and third digits, the numbers must be in the order 713. So the license plate number was 67 713.

## Section 4.2

1. a. Since 126 is even, 2 is a factor. $126 \div 2 = 63$. The number 63 is divisible by 3, $63 = 3 \times 21 = 3 \times 7 \times 3$. Or, you may note that $63 = 9 \times 7$. In any case we find that $126 = 2 \times 3 \times 3 \times 7$.

   b. Since 308 is even, 2 is a factor. $308 \div 2 = 154$. The number 154 is also even, so divide by 2 again to get 77. Since $7 \times 11 = 77$, the prime factorization is $208 = 2 \times 2 \times 7 \times 11$.

   c. The number 245 is not even, but it is a multiple of 5. $245 \div 5 = 49$, and $49 = 7 \times 7$. So the prime factorization is $245 = 5 \times 7 \times 7$.

3. For each of these numbers there are several different correct ways to create the factor trees. The final list of prime factors is unique though. Two possible trees are shown for each. The numbers at the ends of the branches make up the prime factorization.

   a.

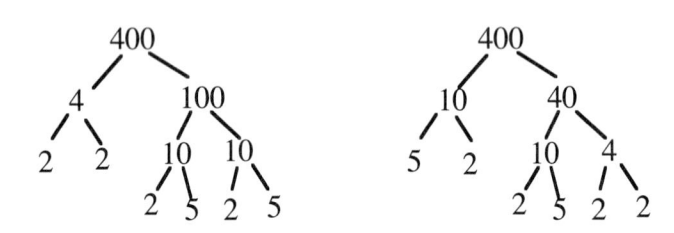

   b.

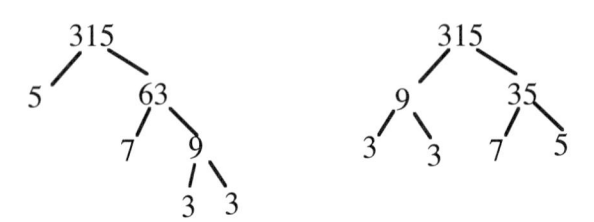

   c.

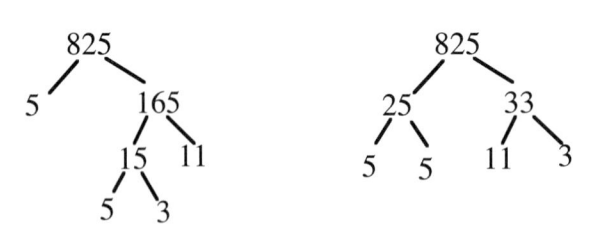

5. Since 1,000,000,000 is equal to $10^{12}$, and 10 is $5 \times 2$, the prime factorization of one billion contains 12 factors of 5 and 12 factors of 2.
   In symbols, $1,000,000,000 = 10^{12} = 5^{12} \times 2^{12}$. The *Fundamental Theorem of Arithmetic* says that the prime factorization of any number is unique. So there can not be any 7's in the prime factorization of one billion. Hence 7 can not be a factor.

## 4.2 Greatest Common Divisor and Least Common Multiple

7. The factors for numbers usually come in pairs. The only exception occurs when we have a square number. We need to check for factors up to the square root of the number. A second way to find all of the factors of a number is to find the prime factorization of the number and then also look at all possible combinations of two or more of the prime factors. The first technique is shown in part a and the second method is shown in b and c.

   a. Factor pairs for 500 include 1 × 500, 2 × 250, 4 × 125, 5 × 100, 10 × 50, and 20 × 25. There are no other factors of 500 less than 23, so the factors of 500 are 1, 2, 4, 5, 10, 20, 25, 50, 100, 125, 250, and 500.

   b. The prime factorization of 231 is 3 × 7 × 11. So in addition to 1, 231, 3, 7, and 11, there are also factors of 21, 33, and 77. The list of factors is 1, 3, 7, 11, 21, 33, 77, and 231.

   c. The prime factorization of 245 is 5 × 7 × 7. So in addition to 1, 245, 5, and 7, there are also factors of 35 and 49. The list of factors is 1, 5, 7, 35, 49, and 245.

9. a. The factors of 30 are 1, 2, 3, 5, 6, 10, 15, and 30.
      The factors of 40 are 1, 2, 4, 5, 8, 10, 20, and 40. The common factors are 1, 2, 5, 10.

   b. The factors of 15 are 1, 3, 5, and 15.
      The factors of 22 are 1, 2, 11, and 22. The only common factor is one.

   c. The factors of 14 are 1, 2, 7, and 14. The factors of 56 are 1, 2, 4, 7, 8, 14, 28, and 56. The common factors are 1, 2, 7, 14. *[Why were all of the factors of 14 common factors?]*

11. We can find the greatest common factor by listing all of the factors of the numbers and finding the largest of the ones they have in common. If the numbers are large or have many factors then it is more efficient to look at the prime factorizations.

    a. To find the GCF(280, 168) we look at the prime factorizations of 280 and 168. 280 = 2 × 2 × 2 × 5 × 7 and 168 = 2 × 2 × 2 × 3 × 7. The prime factors that they have in common are 2, 2, 2, and 7. Since 2 × 2 × 2 × 7 = 56, the GCF(280, 168) = 56.

    b. To find the GCF(12, 15, 125) we look at the prime factorizations of 12, 15, and 125. 12 = 2 × 2 × 3, 15 = 3 × 5 and 125 = 5 × 5 × 5. The three prime factorizations have no common factors that are present in all three. Since 1 is a factor of any whole number the answer is GCF(12, 15, 125) = 1.

    c. To find the GCF(198, 65) we can look at the prime factorizations. 198 = 2 × 3 × 3 × 11. 165 = 3 × 5 × 11. There are common factors of 3 and 11. GCF(198, 65) = 3 × 11 = 33.

13. a. Multiples of 4 are 4, 8, 12, 16, 20, 24, 28, ... Multiples of 14 are 14, 28, 42, 56, ... The first number in common in both lists is 28. The next one will be 28 × 2 = 56. The first 5 common multiples of 4 and 14 are 28, 56, 84, 112, 140.

　　b. Multiples of 6 are 6, 12, 18, 24, 30, ... Multiples of 8 are 8, 16, 24, 32, ... The first five common multiples of 6 and 8 are 24, 48, 72, 96, 120.

　　c. The numbers 12 and 17 are relatively prime. Their first common multiple is 12 × 17 = 204. The first five common multiples of 12 and 17 are 204, 408, 612, 816, 1020.

15. a. To find the LCM(22, 56) we can either look at lists of multiples of both numbers, or we can determine the LCM by using the prime factorizations of 22 and 56. Since the numbers are fairly large we will use the factorizations. 22 = 2 × 11 and 56 = 2 × 2 × 2 × 7. The LCM must include each factor from each of the numbers, but we don't need to repeat factors that appear in both numbers. In this case we need three factors of 2 and a 7 from 56 and we need the factor of 11 from 22, but we don't need another factor of 2 because it is already taken care of by using the 2's in 56. So LCM(22, 56) = 2 × 2 × 2 × 7 × 11. This is 616.

　　b. We will again look at prime factorizations. 6 = 2 × 3. 38 = 2 × 19. 16 = 2 × 2 × 2 × 2. We need to use all four factors of 2 that make up 16 to get a multiple of 16. In addition we need a factor of 3 and a factor of 19 to ensure that we have a multiple of both 6 and 38. So LCM(6, 38, 16) = 2 × 2 × 2 × 2 × 3 × 19 = 912.

　　c. We will again look at prime factorizations. 30 = 2 × 3 × 5 and 42 = 2 × 3 × 7. LCM(30, 42) = 2 × 3 × 5 × 7 = 210.

17. a. The diagram shows that 36 is a common factor of 6 and 9. Rods of length 6 can be placed in a row to give the same length as rods of length 9. The least common multiple is seen half way across the diagram where the rods first match at 18.

　　b. The diagram shows that 6 is a common factor of both 36 and 24. Since both 36 and 24 contain an even number of 6's, it is also true that 12 is a common factor of 36 and 24.

19. a. The length of an all-orange train is a multiple of 10. All-brown trains have lengths 8, 16, 24, 32, 40, 48, ... The first time they would match would be a train of length 40, the next time would be a train of length 80, then 120, etc. All-brown trains of length 40, 80, 120, etc. always contain a multiple of 5 brown rods. Five brown rods is the same length as 4 orange rods.

　　b. The shortest possible match for orange and brown is 40. The brown train has 5 rods.

21. a. To answer this question we need to know the LCM(7, 15, 81). We will look at their prime factorizations. 7 is prime. 15 = 3 × 5. 81 = 3 × 3 × 3 × 3. The least common multiple is 7 × 3 × 5 × 3 × 3 × 3 = 2835. If these three lights flash together, they will next flash together after 2835 seconds. There are 3600 seconds in an hour, so this is less than an hour. *[How many minutes is it?]*

b. During the 2835 seconds there are 2835 ÷ 7 = 405 births, 2835 ÷ 15 = 189 deaths, and 2835 ÷ 81 = 35 immigrants. This is a net gain of 405 − 189 + 35 = 251 in population.

23. If we find a common factor of 126, 180, and 198, then the students in each grade could be numbered off by that number, with no students left over. To find the largest possible number of teams, we will find the GCF(126,180,198). First look at the prime factorizations. 126 = 2 × 3 × 3 × 7. 180 = 2 × 2 × 3 × 3 × 5. 198 = 2 × 3 × 3 × 11. Each of the numbers has at least one factor of 2 and two factors of 3. The GCF is 2 × 3 × 3 or 18. So the students in each grade can be put into at most 18 different groups, forming 18 teams. Each team will have 126/18 = 7 $3^{rd}$ graders, 180/18 = 10 $4^{th}$ graders, and 198/18 = 11 $5^{th}$ graders, for a total of 28 students per team.
*[What are other possible numbers of teams and team sizes?]*

25. a. Since we are not mixing cookies in piles, and there must be the same number of cookies in each pile, we need the number of cookies in a pile to be a divisor of 300 and of 264. For the largest possible number of cookies in a pile we need the GCF(300, 264). Prime factorizations are: 300 = 2 × 2 × 3 × 5 × 5 and 264 = 2 × 2 × 2 × 3 × 11. The greatest common factor is 2 × 2 × 3 = 12. There will be 12 cookies in each pile.

   b. There are 12 cookies per pile and 300 chocolate chip cookies, so there will be 300 ÷ 12 = 25 piles of chocolate chip cookies.

   c. There are 12 cookies per pile and 264 peanut butter cookies, so there will be 264 ÷ 12 = 22 piles of peanut butter cookies.

27. There are several ways to solve this problem. One way is to list times when the various cuckoos come out. The 10 minute cuckoo comes out at 5:10, 5:20, 5:30, 5:40, 5:50, etc.
The 15 minute cuckoo will sing with the 10 minute cuckoo at 5:30, 6:00, 6:30, etc.
The 25 minute cuckoo (an unusual bird!) will come out at 5:25, 5:50, 6:15, 6:40, 7:05, 7:30.
They will all cuckoo together again at 7:30.
Another way to solve this problem is to look for the LCM(10, 15, 25). Using the prime factorizations 10 = 2 × 5, 15 = 3 × 5, and 25 = 5 × 5 we get 2 × 3 × 5 × 5 = 150 as the least common multiple. So the time when all the cuckoos come out together next is 150 minutes after 5:00, or 7:30.

29. During the first six days of the membership, day 5 is the only one in which neither sister uses the club. They both go on days 1 and 6, Cindy goes alone on days 2 and 4, and Nicole goes alone on day 3. Look at the rest of the 180 days in six day blocks. For example, let's look at days 7 through 12. Cindy goes on days 8, 10, and 12. Nicole goes on days 9 and 12. Neither sister goes on day 7 or 11. During days 13 through 18, neither sister goes on days 13 and 17. This pattern repeats every 6 days.

There are 30 sets of six days in the 180 days. During the first set of six days they only miss one day. During the other 29 sets of 6 days they miss two days each time. So there are a total of 29 times 2 plus 1, or 59 days in which neither sister uses the club.

31. To find the number of zeros at the right end of a number we need to know how many factors of 10 the number has. A factor of 10 consists of a factor of 2 and a factor of 5. To find the number of factors of 10 in the product 1 × 2 × 3 × 4 × ... × 98 × 99 × 100 we need to count factors of 2 and factors of 5. But to get a factor of 10 we need both a five and a 2. Since every even number contains at least one factor of 2, there are many more factors of 2 in 1 × 2 × 3 × 4 × ... × 98 × 99 × 100 than factors of 5. So we just need to count factors of 5. There are 20 multiples of 5 included in the product, namely 5, 10, 15, ... 95, 100. These all contain at least one factor of 5, but 25, 50, 75, and 100 also contain a second factor of 5. None contain 3 factors of 5 because the first such number is 125. So there are a total of 24 factors of 5 in the product 1 × 2 × 3 × 4 × ... × 98 × 99 × 100. This means that the product also has 24 factors of 10 and ends in 24 zeros.

33. Here is a listing of the first 12 natural numbers with their proper factors and classification.

    | number | factors | classification |
    | --- | --- | --- |
    | 1 | none | deficient |
    | 2 | 1 | deficient |
    | 3 | 1 | deficient |
    | 4 | 1,2 | deficient |
    | 5 | 1 | deficient |
    | 6 | 1,2,3 | perfect |
    | 7 | 1 | deficient |
    | 8 | 1,2,4 | deficient |
    | 9 | 1,3 | deficient |
    | 10 | 1,2,5 | deficient |
    | 11 | 1 | deficient |
    | 12 | 1,2,3,4,6 | abundant |

    a. The first perfect number is 6. The next one is 28. (1 + 2 + 4 + 7 + 14 = 28)

    b. There are more deficient than abundant numbers less than 25. The only abundant numbers less than 25 are 12, 18, 20, and 24.
    *[Do you think that this pattern of few abundant numbers continues?]*

35. a. The GCF(99,105) is 3. So the fraction 99/105 can be replaced by an equivalent fraction by dividing both 99 and 105 by 3. The result after using the SIMP key is 33/35.

    b. The prime factorization of 102 is 2 × 3 × 17, and for 275 it is 5 × 5 × 11. There are no common factors. So 102 and 275 are relatively prime. When the SIMP key is pressed the result will remain 102/275.

37. The number 30030 happens to be the product of the first 6 prime numbers, 2, 3, 5, 7, 11, 13. In order for Selene to conclude from the result on the calculator that 211 is prime she needed to know this. Perhaps Selene found the prime factorization 30030 = 2 × 3 × 5 × 7 × 11 × 13. Then since the calculator says that 211/30030 cannot be simplified, it means that none of 2, 3, 5, 7, 11, or 13 are factors of 211. The next prime number is 17 and the square of 17 is 289, so 17 is not a factor of 211 either and 211 must be prime.

# Chapter 4 Test

1. a. The sum of the digits of 48,025 is 19. Since 3 is not a divisor of 19, it is not a divisor of 48,025 either. The statement is false.

   b. Since 3776 is an even number it is divisible by 2. The statement is true.

   c. The sum of the digits of 7966 is 28. Since 3 is not a divisor of 28, it is not a divisor of 7966 either. If 3 is not a factor of 7966, then neither is 6. The statement that 6 does <u>not</u> divide 7966 is true.

   d. The sum of the digits of 4576 is 22. Since 9 is not a divisor of 22, it is not a divisor of 4576 either. The statement is false.

2. a. 3 | 45   b. 12 | 60   c. 20 | 140   d. 17 | 102

3. a. A linear model to show that 78 is a multiple of 6 shows a line segment that is 78 units long, which is subdivided evenly into 6-unit long segments.

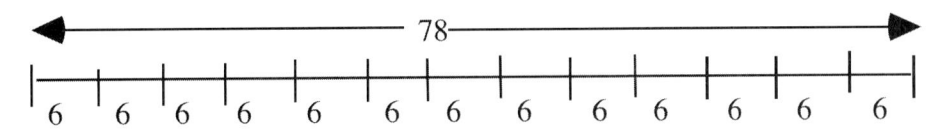

   ```
   ◄─────────────── 78 ───────────────►
   ├──┼──┼──┼──┼──┼──┼──┼──┼──┼──┼──┼──┼──┤
    6  6  6  6  6  6  6  6  6  6  6  6  6
   ```

   b. An area model to show divisibility shows the area divided by one of the dimensions. To show that 6 | 78 we can show a rectangle with area 78 and one dimension 6. If the other dimension is a whole number, then it is divisible. In this case the other dimension is 13.

   ```
   6 │  A = 78  │
     ◄─── 13 ───►
   ```

4. a. A prime number has only one representation as a rectangular array. For example, for the prime number 7 the only array is a 1 by 7 rectangle.

   b. A composite number has more than one representation as a rectangular array. A composite number has other whole number factors besides 1 and itself.

   c. A square number may be represented by any number of arrays, but one of the arrays will be a square, with a single factor. Thus a square number has an odd number of factors.

5. a. The statement is false. A counterexample is 13 because 3 | 3, but 3 does not divide 13.

   b. The statement is true because if a number has 8 as a factor, then its prime factorization contains at least three factors of 2. Only two factors of 2 are needed for divisibility by 4.

   c. False. A counterexample is 31. The sum of the digits is 4, but 31 is not divisible by 2.

   d. False. 9 is not divisible by 6 but it is divisible by 3.

6. a. The number 331 is prime because it is not divisible by 2, 3, 5, 7, 11, 13, or 17. Its square root is less than 19, so that is as far as we need to check.

   b. 351 is composite. The sum of the digits is 9 so it is divisible by both 3 and 9.

   c. 371 is composite because $7 \times 53 = 371$.

7. We can use a factor tree to find the prime factorization. $1836 = 2 \times 2 \times 3 \times 3 \times 3 \times 17$.

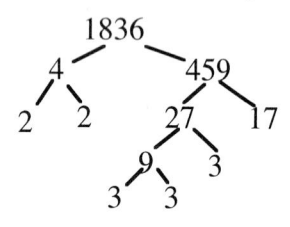

   (Other versions of the factor tree are possible.)
   (But the prime factorization is unique.)

8. The prime factorization of 273 is $3 \times 7 \times 13$. The factors of 273 include all of these and any combinations of them. The factors of 273 are 1, 3, 7, 13, 21, 39, 91, and 273.

9. a. True. If a divides b, then a divides any multiple of b.

   b. False. For a counterexample let a = 2, b = 3, and c = 5. Then 2 | (3 + 5), but 2 does not divide either 3 or 5.

   c. True.

   d. False. For a counterexample let a = 2, b = 3, and c = 4. Then 2 | (3 × 4), but 2 does not divide 3.

10. a. 30 and 40 are both multiples of 10, so 10 is a common factor.
    This means that 1, 2, and 5 are also common factors.
    The four common factors are 1, 2, 5, and 10.

   b. The least common multiple of 15 and 20 is 60. Any multiple of 60 is also a common multiple. Four examples are 60, 120, 180, and 240.

   c. Since both 195 and 255 end in 5, they both have factors of 5. They also are both multiples of 3 because of the sums of the digits. This means that they both have a factor of 15 as well. Four common factors are 1, 3, 5, and 15.

   d. The numbers 13 and 20 are relatively prime, so their least common multiple is their product, which is 260. Any common multiple will be a multiple of 260. The 5 smallest common multiples are 260, 520, 780, 1040, and 1300.

11. a. Since 17 is prime and 30 is not a multiple of 17 they have no common factors (except 1). Hence the GCF(17, 30) = 1. So 17 and 30 are relatively prime.

   b. To find the LCM(14, 22) we can use the prime factorizations of 22 and 56.
        $14 = 2 \times 7$ and $22 = 2 \times 11$.
    The LCM must include each factor from each of the numbers, but we don't need to repeat factors that appear in both numbers. In this case we need to use one factor of 2, one of 7, and one of 11. So the LCM(14, 22) = $2 \times 7 \times 11 = 154$.

   c. To find LCM(12, 210) we use the prime factorizations
        $12 = 2 \times 2 \times 3$ and $210 = 2 \times 3 \times 5 \times 7$.
    For the LCM we need to use two factors of 2, one of 3, one of 5 and one of 7.
    So the LCM(12, 210) = $2 \times 2 \times 3 \times 5 \times 7 = 420$.

   d. $280 = 2 \times 2 \times 2 \times 5 \times 7$ and $165 = 3 \times 5 \times 11$.
    The only common factor is 5, so GCF (280, 165) = 5.

   e. $18 = 2 \times 3 \times 3$   $28 = 2 \times 2 \times 7$   $36 = 2 \times 2 \times 3 \times 3$
    Each has a factor of 2 but no other factors are common to all three.
    GCF(18, 28, 36) = 2.

   f. $6 = 2 \times 3$   $15 = 3 \times 5$   $65 = 5 \times 13$
    To create the LCM we use just enough factors so that each of the above are included.
    So LCM(6, 15, 65) = $2 \times 3 \times 5 \times 13 = 390$.

12. a. The linear model below shows that lengths of 3 strung together and lengths of 8 strung together will first meet again at a length of 24. This is the LCM.

    |—3—|—3—|—3—|—3—|—3—|—3—|—3—|—3—|
    |———8———|———8———|———8———|
    ◄——————— 24 ———————►

    b. The model below shows that 15 and 24 can both be constructed using lengths of 3.

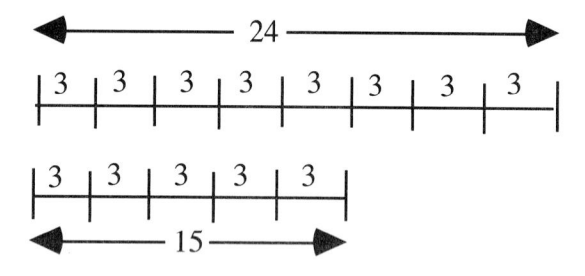

    ◄——————— 24 ———————►
    |—3—|—3—|—3—|—3—|—3—|—3—|—3—|—3—|

    |—3—|—3—|—3—|—3—|—3—|
    ◄——— 15 ———►

13. We need to find the LCM(10, 12). This can be found by listing multiples or by analyzing the prime factorizations. Either way, LCM(10, 12) = 60. The lighthouses will both flash together again after 60 seconds or one minute.

14. One strategy to approach this problem is to look at a simpler problem. Suppose we wanted to count how many whole numbers between 1 and 50 are multiples of either 3 or 7. There are 16 multiples of 3 between 1 and 50. They are the numbers 3, 6, 9, 12, . . . 42, 45, 48. There are 7 multiples of 7, the numbers 7, 14, 21, 28, 35, 42, 49. But the numbers 21 and 42 are in both lists, so we don't want to count them twice. There are 16 + 7 – 2 = 21 multiples of either 3 or 7 between 1 and 50.

    Similarly, there are 999 ÷ 3 = 333 multiples of 3 between 1 and 1000. And there are 994 ÷ 7 = 142 multiples of 7 between 1 and 1000. (1001 is the next multiple of 7.) But the multiples of 21 are counted in both of these lists. There are 987 ÷ 21 = 47 multiples of 21 between 1 and 1000. So there are 333 + 142 – 47 = 428 whole numbers between 1 and 1000 that are multiples of either 3 or 7.

15. Assuming Mike has a saw which will not waste any wood, he is looking for the greatest common factor of 28 and 70. By noting that 7 is a factor of both numbers we can then check for a greater common factor by checking multiples of 7. The greatest common factor is 14.
    This can also be seen in their prime factorizations. $28 = 2 \times 2 \times 7$ and $70 = 2 \times 5 \times 7$. The largest length piece Mike can cut without wasting any wood is 14 inches.

## Chapter 5 Integers and Fractions
## Section 5.1

1. The warmest of these temperatures is ⁻17°. Since all of the temperatures in this list are negative, the one closest to zero is the highest.

3. a. ⁻3 + 1 = ⁻2. Since ⁻3 is to the left of ⁻2 on the number line, ⁻3 < ⁻2.

   b. ⁻14 + 17 = 3. Any positive number is greater than any negative number, so 3 > ⁻14.

   c. ⁻7 + 8 = 1. Any negative number is less than any positive number, so ⁻7 < 1.

5. Arrows in the diagram below connect integers to their opposites.

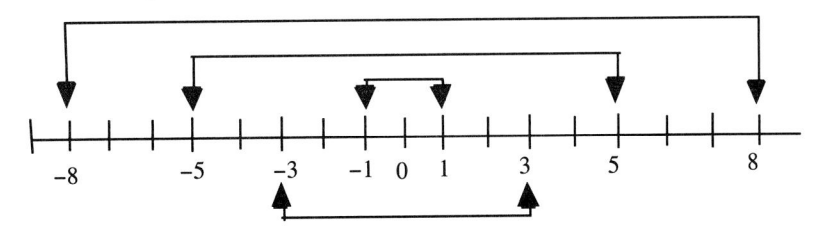

7. a. If the temperature warmed up by 8 degrees, then we need to add 8 to ⁻15.  8 + ⁻15 = ⁻7. The temperature had warmed up to ⁻7° by noontime.

   b. The second quarter trade balance was less than the third quarter. We need to subtract 9 from ⁻23.  ⁻23 – 9 = ⁻32. The second quarter trade balance was ⁻32 billion dollars.

9. In the diagrams below a letter B represents a black chip and R represents a red chip. There are many different ways to represent each number. Three possibilities are shown for each.

   a. To represent ⁻7 there are seven more reds than blacks.

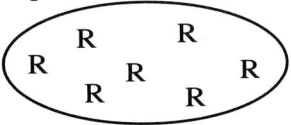

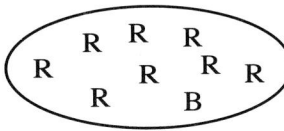

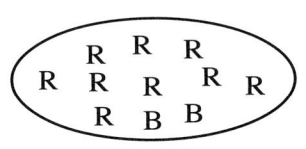

   b. To represent 0 we need an equal number of reds and blacks.

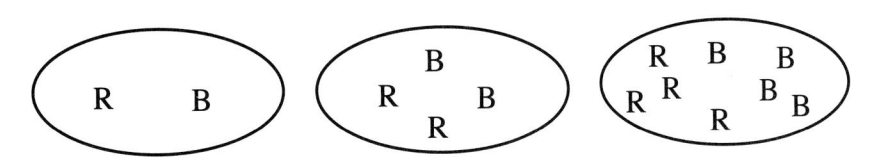

   c. To represent 3 we need three more blacks than reds.

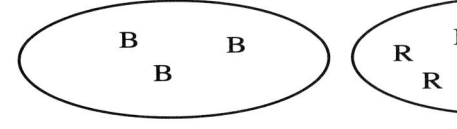

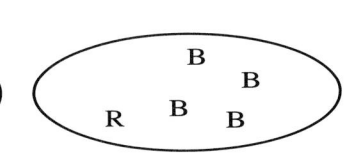

11. a. To illustrate ⁻15 ÷ 5 we show a set of 15 red chips partitioned into 5 groups. Each group has a value of ⁻3, so ⁻15 ÷ 5 = ⁻3.

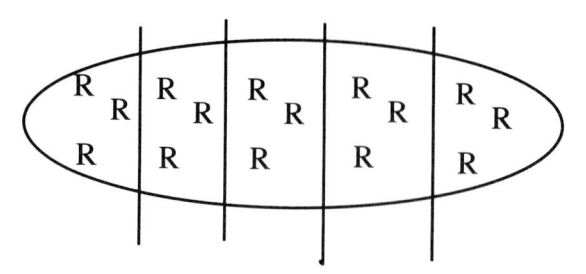

   b. To illustrate ⁻12 ÷ ⁻4 we can show a set of 12 chips grouped into groups of ⁻4. Since there are three groups of ⁻4 in ⁻12, we see that ⁻12 ÷ ⁻4 = 3. Another way to illustrate ⁻12 ÷ ⁻4 would be to show repeated subtraction of ⁻4 from ⁻12.

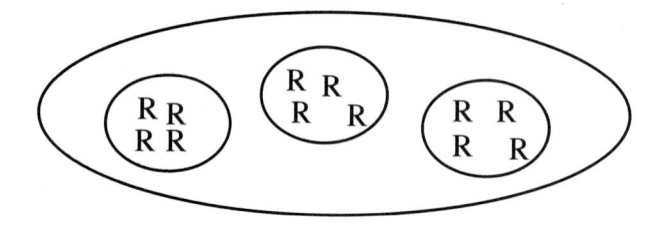

   c. To illustrate 3 × ⁻5 we show three groups of ⁻5 combined together. The result is ⁻15.

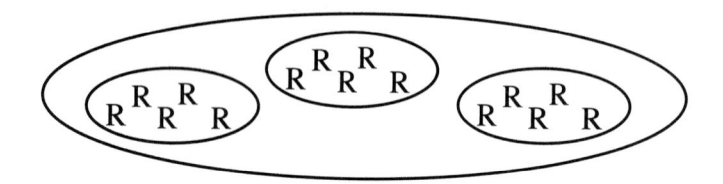

13. In order to take away 5 we need to have at least 5 black chips in the set. In order to show 2 − 5 we want to start with a set that has a value of positive 2, but contains at least 5 black chips. The diagram below shows that 2 − 5 = ⁻3.

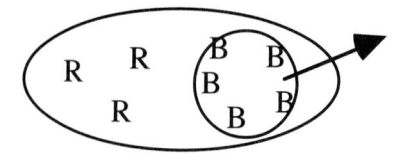

15. a. Starting from the temperature now, we can think of two hours ago as ⁻2 hours and a decrease of 6 degrees per hour as ⁻6. Since ⁻2 × ⁻6 = 12, the temperature was 12 degrees higher two hours ago. It was 30° + 12° = 42°. We also used ⁻2 × ⁻6 = 12.

   b. This time the temperature has been increasing 4° each hour. To find what it was 6 hours ago we add 4 × ⁻6 to the current temperature of 12°. The result is ⁻12°. We used the multiplication fact of 4 × ⁻6 = ⁻24.

## 5.1 Integers

17. To illustrate 6 + ⁻5, we have an arrow starting at 0 and going six spaces to the right, then five spaces to the left. The result is 1.

    To illustrate ⁻4 + 9, we have an arrow starting at 0 and going four spaces to the left, then nine spaces to the right. The result is 5. See the diagrams below.

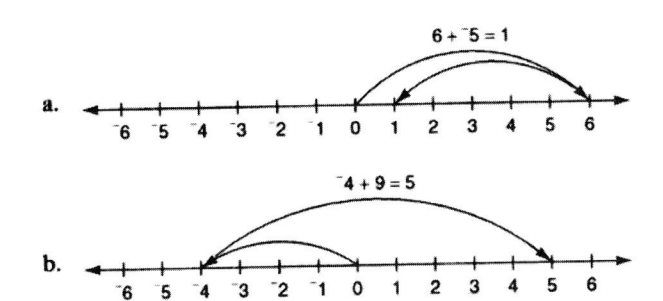

    a.

    $$6 + {}^-5 = 1$$

    b.

    $${}^-4 + 9 = 5$$

19. In each of these problems recall that in multiplying or dividing two numbers, if both signs are the same then the result is positive. If the signs are different then the result is negative.

    a. 14 × ⁻23 = ⁻322.

    b. ⁻156 ÷ 13 = ⁻12.

    c. ⁻278 × ⁻46 = 12,788.

    d. ⁻1431 ÷ ⁻53 = 27.

21. a. 4 + ⁻**14** = ⁻10  because 14 red chips matched with 4 black chips gives a value of 10 red.

    b. 6 – ⁻**4** = 10. Subtracting negative four is the same as adding positive four.

    c. ⁻6 × **2** = ⁻12.

    d. ⁻3 + **3** = 0. Adding opposites results in zero.

23. a. This is an example of the Associative property for multiplication. The only difference between the two sides of the equation is the grouping of the last three numbers:
    2 × (⁻6 × ⁻5) was rewritten as (2 × ⁻6) × ⁻5.

    b. This is an example of the Commutative property for multiplication. The only difference between the two sides of the equation is the change in order of multiplication from 16 × ⁻5 to ⁻5 × 16.

25. a. Subtraction is closed on the set of integers. Whenever an integer is subtracted from an integer the result is always an integer.

    b. Division is not closed on the set of integers. We can find two integers whose quotient is not an integer. For example, 6 ÷ 12 = 0.5.

27. a. There are two pairs of compatible numbers here. 17 + 13 = 30 and ⁻125 + ⁻25 = ⁻150. So the entire sum is 30 + ⁻150 = ⁻120. Other approaches will also work.

94            Chapter 5 Integers and Fractions

b. We can think of $^-298$ as $^-300 + 2$. Then the sum is $700 + (^-300 + 2) + 135$. By the Associative property this is equal to $(700 + {^-300}) + 2 + 135 = 400 + 137 = 537$. Again there are other possible mental techniques.

29. a. For equal products we multiply one number by the same amount that we divide the other. We can divide 24 by 4 and multiply $^-25$ by 4. Then $24 \times {^-25} = 6 \times {^-100} = {^-600}$.

    b. To get an equal quotient we can either divide both numbers by the same quantity. (Remember that here the dividend is the area of a rectangle and the divisor is one of the dimensions. In multiplication we start with the two dimensions.)
We can divide both $^-90$ and 18 by 9 to get $^-90 \div 18 = {^-10} \div 2 = {^-5}$.

    c. Divide $^-28$ by 2 and multiply 5 by 2. Then $^-28 \times 5 = {^-14} \times 10 = {^-140}$.

    d. Divide both numbers by 4. Then $400 \div {^-16} = 100 \div {^-4} = {^-25}$.

31. a. Rounding each number to its leading digit we get $80 + {^-40} + 20 + {^-40} = 20$.

    b. Rounding each number to its leading digit we get $^-20 + 50 + {^-50} + {^-80} = {^-100}$.

(It is helpful to look for compatible numbers after rounding on these.)

33. a. We can round $^-241$ to $^-240$. Then we have $^-240 \div 60 = {^-4}$. Equal quotients are also applicable here, since $^-240 \div 60 = {^-24} \div 6$.

    b. $64 \times {^-11} \approx 64 \times {^-10} = {^-640}$. *[For an exact answer we can use the distributive property to say: $64 \times {^-11} = 64 \times ({^-10} + {^-1}) = {^-640} + {^-64} = {^-704}$.]*

35. a. The answer will be positive. We are taking the product of four numbers, two of which are negative. Since there are an even number of negative factors the product is positive.

    b. The answer will be negative. The product of the first three numbers is negative since there is only one negative factor. The product is clearly less than $^-50$, so adding 50 will not change the sign.

37. a. The next three equations are $5 \times {^-1} = {^-5}$, $5 \times {^-2} = {^-10}$, $5 \times {^-3} = {^-15}$. This suggests that positive times negative gives negative.

    b. The next three equations are $^-1 \times 6 = {^-6}$, $^-2 \times 6 = {^-12}$, $^-3 \times 6 = {^-18}$. This suggests that negative times positive gives negative.

39. a. This will perform the subtraction $^-217 - 366 = {^-583}$.

b. This will perform the addition $^-483 + 225 = {}^-258$.

c. This will perform the division $2257 \div {}^-37 = {}^-61$.

d. This will perform the division $^-1974 \div 42 = {}^-47$.

41. a. To add two negative numbers, we add their absolute values and the answer is negative. On the calculator, add $487 + 653$ to get 1140. The answer for $^-487 + {}^-653$ is $^-1140$.

    b. Subtracting a negative number is the same as adding its opposite, a positive. So to perform $360 - {}^-241$ on a calculator we can simply add $360 + 241 = 601$.

    c. When multiplying a positive number by a negative number the answer is always negative. On the calculator find $32 \times 14$. This is 448, so $32 \times {}^-14 = {}^-448$.

    d. When dividing a positive number by a negative number the answer is always negative. On the calculator find $336 \div 16 = 21$. So $336 \div {}^-16 = {}^-21$.

43. a. We are continually adding $^-17$, starting from 0.
       The sequence will be $^-17, {}^-34, {}^-51, {}^-68, {}^-85$.

    b. We are continually multiplying by $^-26$, starting from 2.
       The sequence will be $^-52, 1352, {}^-35152, 913952, {}^-23762752$.

45. a. This is $^-15 - 6 = {}^-21$.

    b. This is $^-15 - {}^-6 = {}^-15 + 6 = {}^-9$.

    c. This might be $^-15 - {}^-6 = {}^-15 + 6 = {}^-9$.

    Part b. computes $^-15 - {}^-6$. Part c. might compute the correct answer, depending on the calculator.

47. a. To find the time elapsed between $^-80$ days and $^-29$ days we can subtract $^-29 - {}^-80$. The result is $^-29 - {}^-80 = {}^-29 + 80 = 51$ days. One can also think of this as subtracting $80 - 29$.

    b. There are 17 days elapsed between $^-18$ days and $^-1$ day. An additional $24 - 7 = 17$ hours have also elapsed. The total elapsed time is 17 days and 17 hours (or, 425 hours).

    c. The closest approach was at time 0. Thirty-six hours after time $^-7$ hours would be the same as 29 hours after time 0. A number line may be useful for visualizing this one.

49. Since the only day for which we are given the exact temperature is the last day, Friday, working backward may be a good strategy. Friday's temperature of ⁻14° was half of Thursday's temperature. So Thursday's temperature was twice ⁻14°, or ⁻28°. Thursday's temp. was 4° higher than Wednesday's, so Wednesday's temp. was ⁻28° − 4° = ⁻32°. Wednesday's temp. was twice as low as Tuesday's, so Tuesday's temperature was ⁻32° ÷ 2 = ⁻16°. Tuesday was 8 degrees colder than Monday, so Monday's temperature was ⁻16° + 8° = ⁻8°.

51. We can only use three plus or minus signs. To reach a sum of 50 we will probably use three two-digit numbers and one single digit number. To convince yourself of this try setting up the problem using a three digit number. After trying several combinations of positive and negative numbers, we find that 12 − 34 + 5 + 67 = 50.

## Section 5.2

1. a. There are numerous fractions to choose from in the collage. Some examples with different denominators include these four: 1/2, 1/3, 3/4, 1/8.

   b. The common denominator for 1/2 and 1/3 is 6 because it is the LCM of 2 and 3.
   The common denominator for 1/2 and 1/4 is 4 because it is the LCM of 2 and 4.
   The common denominator for 1/2 and 3/4 is 4 because it is the LCM of 2 and 4.
   The common denominator for 1/2 and 1/8 is 8 because it is the LCM of 2 and 8.
   The common denominator for 1/3 and 3/4 is 12 because it is the LCM of 3 and 4.
   The common denominator for 1/3 and 1/8 is 24 because it is the LCM of 3 and 8.
   The common denominator for 3/4 and 1/8 is 8 because it is the LCM of 4 and 8.
   Other choices of fractions may give other common denominators.

3. a. In this figure four of the nine squares are shaded. The shaded part is 4/9 and the unshaded part is 5/9.

   b. The shaded part is 3/7 and the unshaded part is 4/7.

   c. If the entire circle were divided into the small size pieces there would be twelve parts. Five of these parts would be shaded. The shaded part is 5/12. The unshaded part is 7/12.

5. a. Using the sharing concept we will share the three circles among two equal sets. Each set has one and a half circles. The diagram on the left below shows $3 \div 2 = 1\frac{1}{2}$.

   b. Sharing two circles among four sets we get one half circle in each set. The diagram above on the right shows $2 \div 4 = \frac{1}{2}$.

## 5.2 Introduction to Fractions

7. a. The dark green rod is three times the length of the red rod. There are four red rods in a brown rod, so the dark green rod is 3/4 the length of the brown rod. If the dark green rod represents 3/4, then the brown rod is the unit.

   b. The yellow rod is the length of five white rods and the black is seven white rods. The yellow rod represents 5/7 when the black rod is the unit rod.

9. a. To show that 1/3 = 4/12 we group 4 of the 12 dots together and see that it is also one part out of three equal parts.

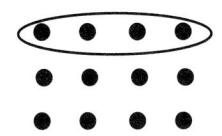

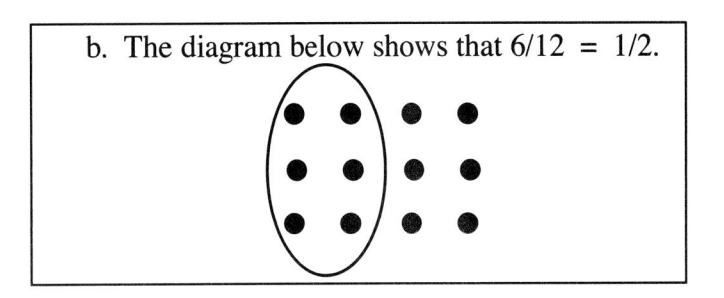

   b. The diagram below shows that 6/12 = 1/2.

   c. This diagram shows 2/3 = 8/12.

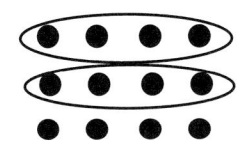

   d. This diagram shows 9/12 = 3/4.

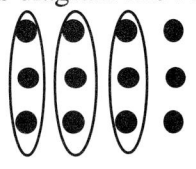

11. a. The missing number is 28. Since 3 × 4 = 12, we need to also multiply 7 by 4 to get an equivalent fraction.

    b. The missing number is 5. 8 is 1/5 of 40 and 5 is 1/5 of 25.

    c. The missing number is 25. We multiply 6 by 5 to get 30, so we also multiply 5 by 5.

    d. The missing number is 4. 9 is 1/3 of 27 and 4 is 1/3 of 12.

13. a. Since 4 and 18 are both multiples of two, we can write an equivalent fraction by dividing both numerator and denominator by two. 4/18 = 2/9.

    b. The GCD of 12 and 27 is 3. The fraction 12/27 in lowest terms is 4/9.

    c. The GCD of 4 and 12 is 4. The fraction 4/12 in lowest terms is 1/3.

    d. The GCD of ⁻16 and 24 is 8. The fraction ⁻16/24 in lowest terms is 2/3.

15. a. Splitting each of the tenths into two equal parts we see that 7/10 = 14/20.

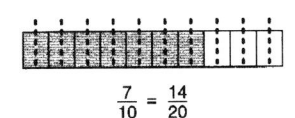

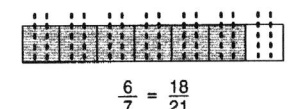

    $$\frac{7}{10} = \frac{14}{20} \qquad \frac{6}{7} = \frac{18}{21}$$

    b. Splitting each of the sevenths into three equal parts we see that 6/7 = 18/21.

17. a. To show that 9/12 = 3/4 we put the nine shaded parts into three equal groups.

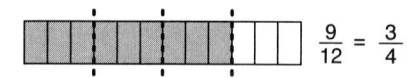

$$\frac{9}{12} = \frac{3}{4}$$

b. To show that 4/6 = 2/3 we put the four shaded parts into two equal groups.

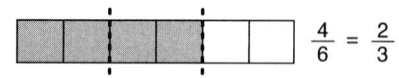

$$\frac{4}{6} = \frac{2}{3}$$

19. a. $\frac{2}{3} = \frac{10}{15}$  b. $\frac{1}{6} = \frac{2}{12}$

$\frac{4}{5} = \frac{12}{15}$     $\frac{-7}{12} = \frac{-7}{12}$

In part a. we use 15 as the common denominator because it is the LCM of 3 and 5. We then form equivalent fractions using 15$^{ths}$. In part b we can use 12$^{ths}$ because both 6 and 12 are divisors of 12.

21. a. $\frac{3}{7} < \frac{5}{9}$ because 3 is less than half of 7, but 5 is more than half of 9.

b. $\frac{1}{4} > \frac{1}{6}$ because fourths are larger pieces than sixths.

c. $\frac{-5}{6} > \frac{-7}{8}$ because $\frac{-5}{6}$ is to the right of $\frac{-7}{8}$ on the number line.

23. a. One way to find a fraction between 1/10 and 1/20 is to write equivalent fractions with larger denominators. Since 1/10 = 4/40 and 1/20 = 2/40, the fraction 3/40 is between 1/10 and 1/20. (There are many other correct answers, such as 4/41 or 7/80 or 9/100.)

b. 1/2 = 8/16 and 5/8 = 10/16, so 9/16 is between 1/2 and 5/8.

25. a. Since 3/3 is one whole, 5/3 is one whole and two more thirds. $5/3 = 1\frac{2}{3}$.

b. Eight parts out of eight is one whole so 8/8 = 1.

c. There are 6/6 in each one, so 24/6 = 4. Then $25/6 = 4\frac{1}{6}$.

27. a. One is 4/4, so we have 4/4 + 3/4, or 7/4.

b. There are 10 fifths in 2, so $-2\frac{1}{5} = -11/5$.

c. There are 12 thirds in 4, so $4\frac{2}{3}$ contains 14 thirds.   $4\frac{2}{3} = 14/3$.

29. The marks on the line in the diagram below divide it into fourths. The improper fractions below the line indicate the number of fourths. The mixed number equivalent for 5/4 is one and one-fourth and for 7/4 it is one and three-fourths.

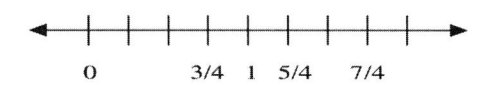

```
◄—+—+—+—+—+—+—+—+—►
   0      3/4  1  5/4  7/4
```

31. a. The drawing below shows fraction bars for 1/12 and 1/20. Since twentieths are smaller than twelfths, 1/12 > 1/20.

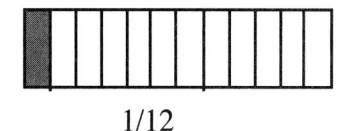

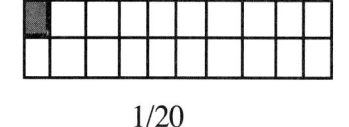

    1/12                1/20

   b. A drawing similar to the one above will show that 7/8 < 9/10. Since 1/10 is smaller than 1/8, the fraction 9/10 is closer to 1 than is 7/8.

   c. Comparing both fractions to 1/2, we see that 5/12 < 1/2 and 6/11 > 1/2. So 5/12 < 6/11.

33. a. 4/10 < 1/2, so the fraction 4/10 is closer to 0 than 1. To the nearest integer it rounds to 0.

   b. Since 1/3 < 1/2, the mixed number $1\frac{1}{3}$ is closer to 1 than 2. It rounds to 1.

   c. $3\frac{1}{2}$ is exactly halfway between 3 and 4. By convention we round it to 4.

35. If we are dividing by the smallest common factor (other than 1) we should first check both numbers for divisibility by 2, then 3, then 5, etc. In the case of the fraction 60/630, the calculator will first divide both numerator and denominator by 2, since both are even. The result in screen b is 30/315. Since 315 is odd, the calculator can not further simplify with factors of 2. But both 30 and 315 are multiples of 3. Screen c shows the result 10/105. The only common factor remaining is 5. Screen d will show 2/21.

37. a. The improper fraction 88/32 is between 2 and 3 because 2 × 32 = 64 and 3 × 32 = 96. Since 88 − 64 = 24, there are 24/32 more than 2 in 88/32. Dividing both 24 and 32 by 8 we can simplify 24/32 to 3/4. So, writing 88/32 as a mixed number in lowest terms the calculator will give $2\frac{3}{4}$.

   b. The improper fraction 154/70 is between 2 and 3 because 2 × 70 = 140 and 3 × 70 = 210. Since 154 − 140 = 14, there are 14/70 more than 2 in 154/70. Dividing both 14 and 70 by 4 we can simplify 14/70 to 1/5. So, writing 154/70 as a mixed number in lowest terms the calculator will give $2\frac{1}{5}$.

100                                Chapter 5   Integers and Fractions

39. A furlong is 1/8 mile, so 6 furlongs is 6/8 or 3/4 mile.  If a horse runs a mile in one and a half minutes, then at the same rate it will run 3/4 of a furlong in 3/4 of the time.  We can get the answer by multiplying 3/4 by one and a half, or we can find 3/4 of a minute and 3/4 of 30 seconds and add them together.  Using the last method we get 45 seconds plus 22.5 seconds.  So the horse will run 6 furlongs in 67.5 seconds, or 1 minute and 7.5 seconds.

41. Since 3/8 is equivalent to 6/16 you should use the 6/16 inch drill bit.

43. The fraction 1/50 is smaller than 1/20.  So there is more iron than magnesium in the earth's crust.

45. Of the 1040 grams in a liter of milk, 40 grams are protein.  So the fraction of the milk's weight that is protein is 40/1040.  This is equivalent to 1/26.

47. If the camera is not getting quite enough light, then a slower shutter speed is desired.  Since 250 means 1/250 of a second, the next slower setting is 125, which means 1/125 of a second.  1/125 of a second is twice as long as 1/250 of a second.

49. a. The recipe calls for $1\frac{2}{3}$ cups of skim milk and we have $1\frac{3}{4}$ cups.  Since 2/3 = 8/12 and 3/4 = 9/12, we have enough.

    b. The recipe calls for 1/3 cup of orange juice and we have 1/4 cup.  Since 1/3 = 4/12 and 1/4 = 3/12, we do not have enough.

51. a. 1/3 is less than 1/2, so this is not enough.

    b. 3/8 cup is less than 1/2 cup.  We are short by 1/8 cup of sugar.

53. In each of the first three fractions in the list, 2/5, 3/8, and 1/3, the denominator is the second Fibonacci number following the Fibonacci number in the numerator.  The second Fibonacci number following 5 is 13.  So for the almond, the phyllotaxis fraction is 5/13.  For the elm it is 1/2 and for the willow it is 3/8.

## Section 5.3

1. a. The actual diameter is 10 times the diameter of the model.  10 × 3.6 inches = 36 inches.

   b. One tenth of 10 feet is 1 foot.

3. a. If the fox's height in the picture is $1\frac{1}{4}$ inch and the scale is 1/12, then the height of the life-size fox is 5/4 × 12 = 15 inches.

   b. Since the scale for the eagle is 1/32, if the photo length is 3/4 inch, then the length of the life-size eagle is 3/4 × 32 = 24 inches.

5.

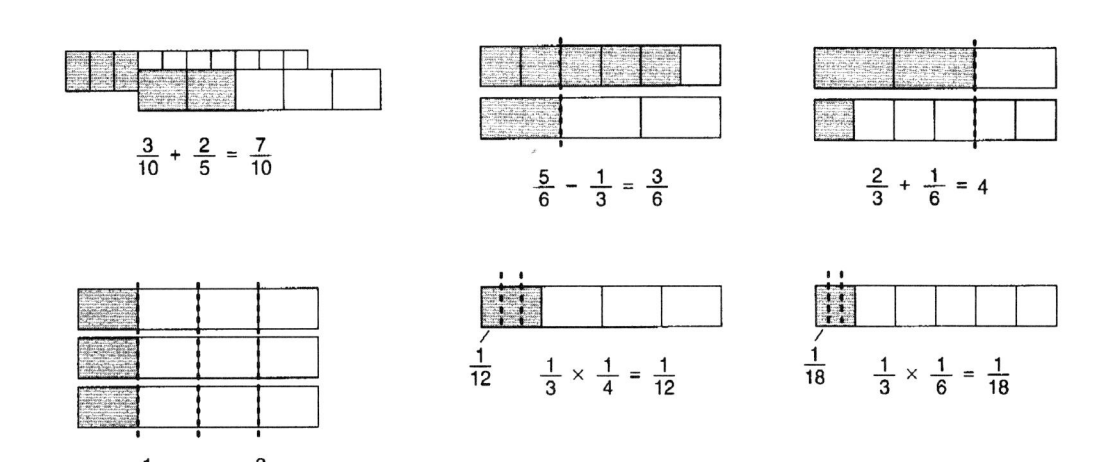

$\frac{3}{10} + \frac{2}{5} = \frac{7}{10}$

$\frac{5}{6} - \frac{1}{3} = \frac{3}{6}$

$\frac{2}{3} + \frac{1}{6} = 4$

$\frac{1}{4} \times 3 = \frac{3}{4}$

$\frac{1}{12}$   $\frac{1}{3} \times \frac{1}{4} = \frac{1}{12}$

$\frac{1}{18}$   $\frac{1}{3} \times \frac{1}{6} = \frac{1}{18}$

7. a. To show 3/8 × 24 we can show 24 dots divided into 8 equal groups and select three of the groups. The diagram below on the left shows that 3/8 × 24 = 9.

3/8 x 24

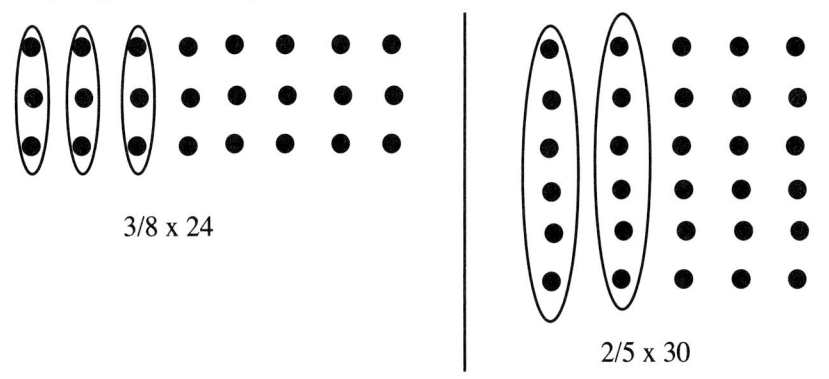

2/5 x 30

b. The diagram above on the right shows that 2/5 × 30 = 12.

9. a. Place the edge of a piece of paper on the eighth's line and mark off the length $1\frac{3}{8}$. Then place the beginning of this marked-off length at the 4/5 point on the fifths' line to approximate the sum $4/5 + 1\frac{3}{8}$. The approximation will be around $2\frac{1}{4}$ or $2\frac{1}{8}$. To calculate the actual sum we find equivalent fractions with common denominators. $4/5 = 32/40$ and $1\frac{3}{8} = 11/8 = 55/40$. The sum is $87/40$ or $2\frac{7}{40}$.

b. Place the edge of a piece of paper on the eighth's line and mark off the length 5/8. Then place the beginning of this marked-off length at the 7/10 point on the tenths' line to approximate the sum 5/8 + 7/10. The approximation will be around $1\frac{3}{10}$. To calculate the actual sum we find equivalent fractions with common denominators. $5/8 = 25/40$ and $7/10 = 28/40$. The sum is $53/40$ or $1\frac{13}{40}$.

11. a. Place the edge of a piece of paper on the eighth's line and mark off the length 5/8. Then place the end of this marked-off length at the $1\frac{4}{5}$ point on the fifths' line and look at the point where the front of the marked length hits. The approximation will be about $1\frac{1}{5}$. To calculate the actual difference we find equivalent fractions with common denominators. 5/8 = 25/40 and $1\frac{4}{5}$ = 9/5 = 72/40. The difference is 72/40 − 25/40 = 47/40 or $1\frac{7}{40}$.

b. Place the edge of a piece of paper on the tenth's line and mark off the length 7/10. Then place the end of this marked-off length at the $1\frac{1}{8}$ point on the eighths' line and look at the point where the front of the marked length hits. The approximation will be about 3/8 or 1/2. To calculate the actual difference we find equivalent fractions with common denominators.
7/10 = 28/40 and $1\frac{1}{8}$ = 9/8 = 45/40. The difference is 45/40 − 28/40 = 17/40.

13. a. 2/3 + 3/4 = 8/12 + 9/12 = 17/12 = $1\frac{5}{12}$.

b. 1/6 + 3/8 = 4/24 + 9/24 = 13/24.

c. 2/3 × 6 = 2/3 × 6/1 = 12/3 = 4. (This one can also be done by thinking of 2/3 of 6.)

d. ⁻3/4 × 2/5 = ⁻6/20 = ⁻3/10.

e. $2\frac{1}{4} + 1\frac{1}{3}$ = 9/4 + 4/3 = 27/12 + 16/12 = 43/12 = $3\frac{7}{12}$. Or, the 1/4 and 1/3 can be added separately to give 7/12. Then add the 2 and the 1 to get $3\frac{7}{12}$.

f. $^-1\frac{5}{6} + 3\frac{1}{2}$ = ⁻11/6 + 7/2 = ⁻11/6 + 21/6 = 10/6 = 5/3 = $1\frac{2}{3}$.

g. $2\frac{1}{4} \times 3\frac{1}{2}$ = 9/4 × 7/2 = 63/8 = $7\frac{7}{8}$.

h. $14\frac{1}{2} \div 2\frac{1}{4}$ = 29/2 ÷ 9/4 = 29/2 × 4/9 = 116/18 = 58/9 = $6\frac{4}{9}$.

i. 7/8 − 1/3 = 21/24 − 8/24 = 13/24.

15. The opposites of 7/8, ⁻4, ⁻1/2, and 10 are ⁻7/8, 4, 1/2, and ⁻10. The opposite of a number can be added to it to obtain 0. So the opposite is the additive inverse.

    The reciprocals of 7/8, ⁻4, ⁻1/2, and 10 are 8/7, ⁻1/4, ⁻2, and 1/10. When a number is multiplied by its reciprocal the result is 1. The reciprocal is the multiplicative inverse.

17. a. Instead of multiplying both numerator and denominator by the same number to obtain an equivalent fraction, it looks as if this person <u>added</u> the same amount to numerator and denominator in an attempt to convert both fractions to a common denominator of 8. Note that adding the same amount to the numerator and denominator does not give equivalent fractions. 1+5 = 6 and 3 + 5 = 8, but 1/3 ≠ 6/8.

    b. The subtraction was attempted without getting a common denominator. Both numerators and denominators were subtracted. *[This is a common type of error. How should a teacher respond to this work by a student?]*

    c. It appears the student here may have "cross-multiplied", taking 4 times 2 and 11 times 1.

    d. The student probably divided the numerators and divided the denominators.
       *[Why doesn't this work?]*

19. a. This is the commutative property for addition because the order of addition is reversed in the addition of 2/9 and 1/3.

    b. This equation shows the use of the multiplicative inverse. Since 2/9 and 9/2 are inverses, their product is 1.

    c. The associative property for addition has been applied here. On the left side the last two fractions are associated, and on the right side the first two are associated.

21. a. Addition is closed for the set of positive fractions, because the sum of any two positive fractions is another positive fraction.

    b. Multiplication is not closed for the set of negative fractions, because the product of two negative fractions can be (in fact is always) a positive fraction.

23. a. The fractions with common denominators are compatible. The sum of $5\frac{1}{2} + 1\frac{1}{2}$ is 7. The sum of $^-2\frac{2}{3}$ and $6\frac{1}{3}$ is $3\frac{2}{3}$. So the sum of all four numbers is $7 + 3\frac{2}{3} = 10\frac{2}{3}$.

    b. First subtract $5\frac{5}{8} - 1\frac{1}{8}$. The result is $4\frac{1}{2}$. Then add $2\frac{1}{2}$ to $4\frac{1}{2}$ to get 7.

c. Since multiplication is commutative and associative we can multiply these four numbers in any order. Since 1/5 of 20 is 4, then 2/5 of 20 is 8. So 2/5 × 20 = 8. Since 1/3 of 12 is 4, then ⁻1/3 × 12 = ⁻4. So the original product is equivalent to 8 × ⁻4 = ⁻32.

d. Because of order of operations we must multiply 7/4 times 24 first. Since 1/4 of 24 is 6, 7/4 of 24 is 7 × 6 = 42. So 7/4 × 24 = 42 and 7/4 × 24 + 8 = 42 + 8 = 50.

25. a. $215\frac{1}{5} - 10\frac{4}{5} = 214\frac{6}{5} - 10\frac{4}{5} = 204\frac{2}{5}$.

   b. $4 \times 9\frac{6}{7} = 4 \times (10 - 1/7) = 4 \times 10 - 4 \times 1/7 = 40 - 4/7 = 39\frac{3}{7}$.

   c. $86\frac{11}{12} + 10\frac{1}{2} = (86\frac{5}{12} + 1/2) + 10\frac{1}{2} = 86\frac{5}{12} + (1/2 + 10\frac{1}{2}) = 86\frac{5}{12} + 11 = 97\frac{5}{12}$.

27. a. To subtract $8 - 3\frac{6}{7}$ we can first add 1/7 to each number to obtain the equal difference of $8\frac{1}{7} - 4 = 4\frac{1}{7}$.

   b. Similarly here, add 1/10 to both numbers before subtracting. Then the difference is $15\frac{4}{10} - 11 = 4\frac{4}{10}$.

   c. To divide 5/8 ÷ 1/3 using equal quotients we can multiply both fractions by 3 to give us 15/8 ÷ 3/3, or 15/8 ÷ 1 = 15/8. (Note that in this problem the same thing happens when we convert the division to multiplying by the inverse.)

29. a. The rounded numbers are 2, 3, 2, and 2, so that the approximate sum is 9.

   b. The rounded numbers are 4, 2, 3, and 5, so that the approximate sum is 14.

31. a. $4\frac{1}{3} \times 6\frac{1}{2} \approx (4 \times 6) + (4 \times 1/2) + (6 \times 1/3) = 24 + 2 + 2 = 28$.

   b. $5\frac{1}{4} \times 8\frac{2}{5} \approx (5 \times 8) + (5 \times 2/5) + (8 \times 1/4) = 40 + 2 + 2 = 44$.

33. a. One estimate can be made by replacing 34 with 35. This is a compatible number because it is divisible by 7. 6/7 of 35 is 30, since 1/7 of 35 is 5.
   Another estimate could be made by rounding 6/7 to 1. Then the estimate is 34.

b. To estimate $9\frac{4}{5} + 5\frac{1}{6}$ we can replace $5\frac{1}{6}$ with $5\frac{1}{5}$. Then the sum is $9 + 5 + 1 = 15$.

Another way to make the estimate would be to round to whole numbers. In this case that is just as accurate.

35. The diagram below shows a fraction bar with 2/3 shaded representing the 2/3 of the earth's surface covered by water. The 1/10 fraction bar shows the portion covered by glaciers. To find the combined amount we can convert both fractions to 30ths. The result is 23/30.

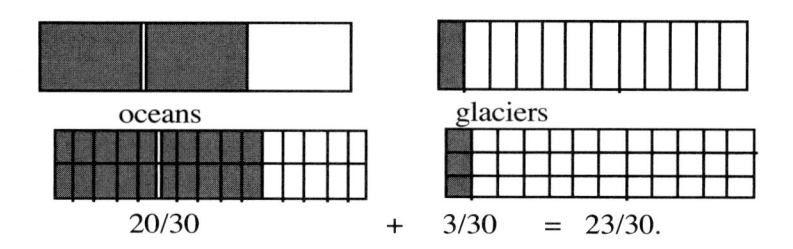

    oceans        glaciers

    20/30    +    3/30    =    23/30.

37. In the diagram below, the entire rectangle represents the 1897 catch of 48 million pounds. One section is shaded to represent 1/6 of the 1897 catch. 1/6 of 48 is 8. There are now 8 million pounds of blue shad caught yearly.

    1897 catch = 48 million pounds

39. The shaded part of the diagram below represents the 1 1/2 inches of rain on Wednesday. Note that 1 1/2 can be thought of as 3/2. On Thursday it rained 1/3 as much as Wednesday, so it rained 1/3 of 3/2 inches. Since there are 3 halves in 3/2 it is easy to see that 1/3 of this amount is one of the halves, or 1/2. It rained 1/2 inch on Thursday.

    ← 3/2 →

    ← → 1/3 of 3/2

41. a. $2/7 + 3/4 = 8/28 + 21/28 = 29/28 = 1\frac{1}{28}$.

    b. $7/8 - 2/3 = 21/24 - 16/24 = 5/24$.

    c. $9/14 \times 4/9 = 4/14 = 2/7$. (The 4/14 was found here by noting that 9/9 = 1.)

    d. $5/6 \div 7/8 = 5/6 \times 8/7 = 40/42 = 20/21$.

106     Chapter 5 *Integers and Fractions*

43. Because of order of operations, the calculator steps in Jan's calculation will produce the correct answer. First 3 is divided by 5 to produce 3/5 or .6, then 80 is divided by .6 to give 400/3 or 133.333... Carl's steps will not produce the correct answer. Using Carl's steps the calculator will first divide 80 by 3, resulting in 80/3 or 26.666... Then this number is divided by 5 to give 80/15 or 5.333...

45. Line 1: 1/3 of $22,000 can be found either by multiplying 22,000 by 1/3 or by dividing 22,000 by 3. Either way, to the nearest cent it is $7333.33.
    Line 2: 1/5 of $12,000 can be found either by multiplying 12,000 by 1/5 or by dividing 12,000 by 5. Either way, to the nearest cent it is $2400.00.
    Line 3: $7333.33 + $2400.00 = $9733.33.
    Line 6: $2318.00 − $1630.00 = $688.00.
    Line 7: $688.00 × 4 = $2752.00
    Line 8: $2318.00 + $2752.00 = $5070.00.

47. a. Since $1\frac{1}{2} \times 4 = 6$, we need to multiply each ingredient by 4. The following amounts are needed: 2 cups non-fat cottage cheese; 2 cups non-fat plain yogurt; $1\frac{1}{3}$ cups low-fat buttermilk; 1 cup Roquefort cheese; 4 teaspoons white pepper.

    b. 1/3 cup + 3/4 cup + 1/2 cup = 4/12 + 9/12 + 6/12 = 19/12 = $1\frac{7}{12}$ cups.
    Since $2 - 1\frac{7}{12} = 5/12$, there will be 5/12 cup unfilled.

49. a. The D string is 8/9 the length of the C string because 8/9 is 8/9 of 1. (8/9 × 1 = 8/9)
    The E is 8/9 the length of the D, because 8/9 × 8/9 = 64/81.
    The F is <u>not</u> 8/9 the length of the E because 64/81 × 8/9 = 512/729. This is not equivalent to 3/4.
    But G is 8/9 of F because 3/4 × 8/9 = 2/3. Similarly, A is 8/9 of G and B is 8/9 of A. But C is <u>not</u> 8/9 of B.

    b. The keys that are not separated by black keys do not have the 8/9 property. The others do.

51. Drawing a diagram may be helpful for solving this problem. Use the given information to make a sketch. The drawing on the left shows the original class, with two-thirds boys. In the second rectangle, enough boys were traded for girls to make the class composition half of each. This required 4 girls, so we know that that part of the rectangle represents 4 students. From the picture we can see that 4 students make up 1/6 of the class, so there are 24 students in the class.

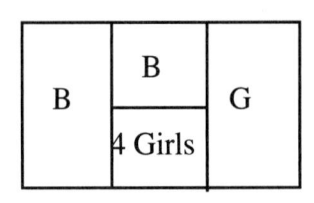

53. Solution 1:
One strategy is to draw a sketch, as in #49. The sections labeled P represent the 4/5 of the discs that Paula removed. The remaining part is divided into 10$^{ths}$, and 7 of the 10$^{ths}$ are labeled S to represent the discs that Sam removed. Since Sam removed 28 discs, each of the small tenths represents 4 discs. So the 10 tenths represent 40 discs. This is still only 1/5 of the box, so the box originally had 40 × 5 = 200 discs, and Paula removed 160 of them.

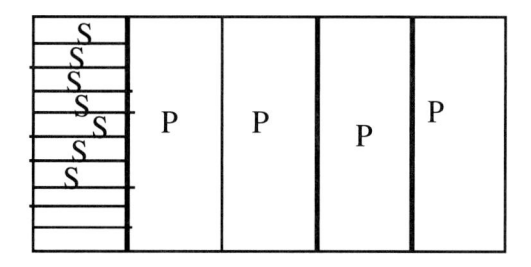

Solution 2:
Sam removed 7/10 of the 1/5 that was remaining, so he removed 7/10 × 1/5 = 7/50 of the discs. Since he removed 28 discs, 28 is 7/50 of the total. So the total is 28 ÷ 7/50.
28 ÷ 7/50 = 28 × 50/7 = 200. Paula removed 4/5 of 200, or 160 discs.

55. In order to form a conjecture we probably want to look at some more examples and see whether we notice a pattern.
  For the fractions 1/2 and 2/3: (1 + 2)/(2 + 3) = 3/5. 3/5 is between 1/2 and 2/3.
  For the fractions 1/4 and 5/7: (1 + 5)/(4 + 7) = 6/11. 6/11 is between 1/4 and 5/7.
  For the fractions 8/21 and 9/21: (8 + 9)/(21 + 21) = 17/42. 17/42 is between 8/21 and 9/21.
  For the fractions 5/2 and 1/3: (5 + 1)/(2 + 3) = 6/5. 6/5 is between 5/2 and 1/3.
For all of these fractions, this method gives a fraction between the original two.
Try some more! You may want to compare their values with a calculator. A reasonable conjecture to make at this point would be to say that for all pairs of fractions this method will result in a fraction between the original two. Can you find any counterexamples?

# Chapter 5 Test

1. In these diagrams, B represents a black chip (positive) and R represents a red chip (negative).
   a. Eight black chips combined with 5 red chips gives a collection with a value of 3 black
      8 + ⁻5 = 3.

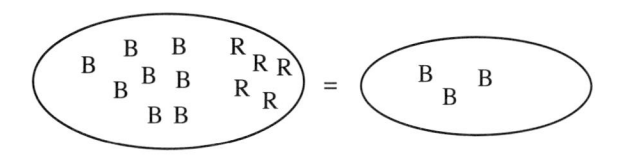

   b. Taking 3 reds away from 7 reds leaves 4 reds. ⁻7 − ⁻3 = ⁻4.

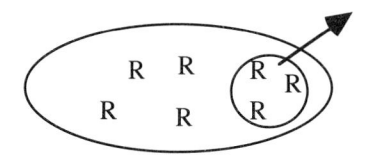

108    Chapter 5  Integers and Fractions

c. Three copies of 4 reds gives 12 reds. $3 \times {}^-4 = {}^-12$.

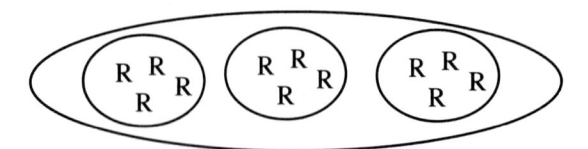

d. If we group 20 red chips into groups of 4 red chips, there will be 5 groups. ${}^-20 \div {}^-4 = 5$.

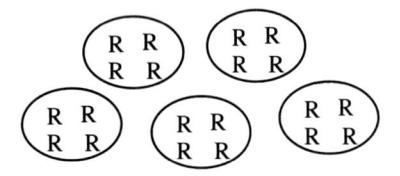

e. Taking 2 blacks away from 6 blacks leaves 4 blacks. $6 - 2 = 4$.

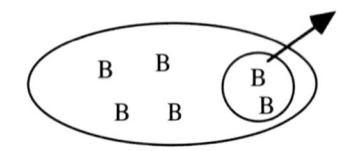

f. Fifteen red chips divided into three groups gives 5 red chips in each group. ${}^-15 \div 3 = {}^-5$.

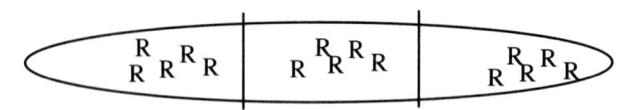

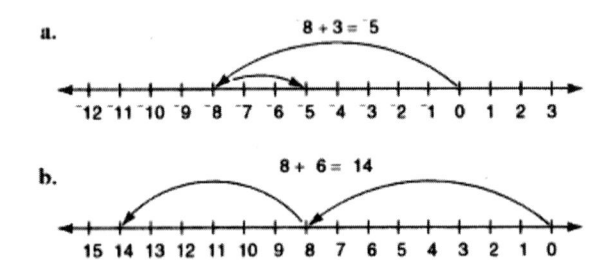

2.
   a. $8 + {}^-3 = 5$

   b. $-8 + {}^-6 = {}^-14$

3. When multiplying or dividing two numbers, if they are both positive or both negative, the answer is positive. If one is negative and one is positive, then the answer is negative.

   a. ${}^-7 \times {}^-6 = 42$   b. $30 \div {}^-5 = {}^-6$   c. $8 \times {}^-10 = {}^-80$   d. ${}^-40 \div {}^-8 = 5$.

4. a. Dividing ${}^-16$ by 4 and multiplying 25 by 4 we get: ${}^-16 \times 25 = {}^-4 \times 100 = {}^-400$.

   b. Dividing both numbers by 8 we get $800 \div {}^-16 = 100 \div {}^-2 = {}^-50$.

   *[Note that we get equal quotients by dividing both numbers, but we get equal products by dividing one factor and multiplying the other one. Why?]*

5.  a. $^-271 \div 30 \approx {^-270} \div 30 = {^-9}$.     270 was chosen because 27 is a multiple of 3.

    b. $1/8 \times 55 \approx 1/8 \times 56 = 7$.

    c. $4 \times 6\frac{1}{5} \approx 4 \times 6 = 24$.

    d. $11 \times {^-34} \approx 10 \times {^-34} = {^-340}$.

6.  a. The six circles are divided into four equal parts. Since each part contains $1\frac{1}{2}$ circles, the quotient is $1\frac{1}{2}$.

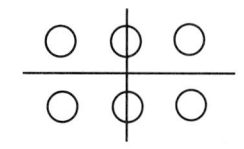

    b. Since 1/3 of fifteen is 5, the product is 5.

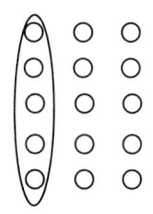

    c. Since 2/3 of 1/5 is 2/15, the product is $2/3 \times 1/5 = 2/15$.

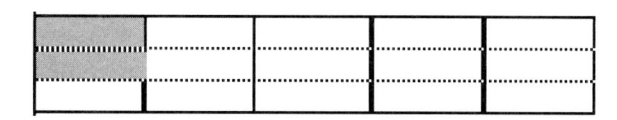

    d. To show that $3/4 = 6/8$ we shade 3/4 and then divide each fourth into two equal parts to get 6/8.

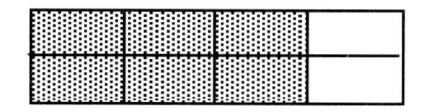

8.  a. $6/11 = 54/99$ and $5/9 = 55/99$, so $6/11 < 5/9$.

    b. $6/11 = 30/55$ and $3/5 = 33/55$, so $6/11 < 3/5$.

    c. Both fractions are between 0 and $^-1/2$. $^-4/9$ is closer to $^-1/2$ than $^-3/7$ is, so $^-4/9 < {^-3/7}$.

9. a. 1/8 > 1/10 because tenths are smaller pieces than eighths are.

  b. 4/7 > 5/12 because 4 is more than half of 7, but 5 is less than half of 12.

  c. 1/2 < 7/12 because 7 is more than half of 12.

  d. 5/6 < 7/8 because sixths are larger than eighths, so one sixth below 1 is less than one eighth below 1.

10. a. We can add 1/6 and 1/3 first to get 1/2. (It is the same as 1/6 + 2/6 = 3/6 = 1/2.) Then add 2 + 4 + 1/2 = $6\frac{1}{2}$.

  b. Subtracting 2/3 − 1/5 gives 2/3 − 1/5 = 10/15 − 3/15 = 7/15. So $5\frac{2}{3} - 1\frac{1}{5} = 4\frac{7}{15}$.

  c. When we add 5/8 and 2/3 the result is more than one, so the sum is more than 8. Adding the fraction parts, we get 5/8 + 2/3 = 15/24 + 16/24 = 31/24 = $1\frac{7}{24}$. Then $6\frac{5}{8} + 1\frac{2}{3} = 8\frac{7}{24}$.

  d. To subtract $4\frac{5}{6}$ from $10\frac{1}{5}$ we need to trade one of the ten units for five more fifths. Then we will have $9\frac{6}{5} - 4\frac{5}{6}$ and we can subtract 5/6 from 6/5 and subtract 4 from 9. Since 6/5 − 5/6 = 36/30 − 25/30 = 11/30, the result is $5\frac{11}{30}$.

11. a. Before converting the mixed numbers to improper fractions it is a good idea to estimate the product. It should be somewhat more than 12 because 6 × 2 = 12. Converting to improper fractions we get 9/4 × 19/3 = 3/4 × 19/1 (because 9/3 = 3/1) = 57/4 = $14\frac{1}{4}$.

  b. Again making an estimate first is a good idea. 8 ÷ 2 = 4. Now convert to improper fractions. So the problem is to divide 44/5 ÷ 17/8. Change to multiplication by inverting the divisor: 44/5 ÷ 17/8 = 44/5 × 8/17 = 352/85 = $4\frac{12}{85}$.

  c. 2/3 × ⁻14 = 2/3 × ⁻14/1 = ⁻28/3 = $⁻9\frac{1}{3}$.

  d. $6 \div 1\frac{1}{2}$ = 6 ÷ 3/2 = 6 × 2/3 = 4. (4 is 2/3 of 6, and there are 4 sets of $1\frac{1}{2}$ in 6.)

12. a. This is an application of the multiplicative inverse. The multiplicative inverse of 3/5 is 5/3, so 3/5 × 5/3 = 1.

   b. This is the distributive property. The multiplication by 3/4 is distributed over the sum of 7 and ⁻1/2.

   c. This is the commutative property of multiplication. The order of the multiplication has been reversed.

   d. The additive inverse is shown here. ⁻4/5 is the additive inverse of 4/5 because their sum is the additive identity zero.

13. a. False. For example, 5 − 3 ≠ 3 − 5.

   b. True. When any two positive integers are multiplied, the result is always in the set of positive integers.

   c. True. When any two positive fractions are multiplied, the result is always in the set of positive fractions.

   d. False. For example, 5 ÷ 2 is not in the set of integers.

   e. False. For example, (7/10 − 4/10) − 2/10 ≠ 7/10 − (4/10 − 2/10).

14. a. The actual shark is 72 times longer than the photo-length. 2 inches times 72 is 144 inches, or 12 feet.

   b. 3/8 inch multiplied by 72 is the same as 3 × 1/8 × 72 = 3 × 9 = 27 inches, or 2 feet and 3 inches.

15. If both pumps are used together, in one hour the new pump will fill 1/5 of the pool and the old pump will fill 1/10 of the pool. Together they fill 1/5 + 1/10 = 3/10 of the pool. In three hours they will have filled 3 × 3/10 = 9/10 of the pool. The remaining 1/10 of the pool will take 1/3 of an hour to fill because they fill 3/10 of the pool in an hour. So it takes $3\frac{1}{3}$ hours or 3 hours and 20 minutes to fill the pool together.

   *[Drawing a diagram may be helpful in understanding this problem.]*

## Chapter 6 Decimals: Rational and Irrational Numbers
### Section 6.1

1. a. The decimal for 1 thousandth of a second is .001.

    b. The decimal for 1 millionth is .000001.

3. a. In Stevin's system the whole number part is written, then a zero in a circle, then the number of tenths, then a one in a circle, then the number of hundredths, then a two in a circle, then the thousandths, followed by a circled three, etc. The number 7.46 looks like:
    $$7⓪4①6②$$

    b. The decimal point may have evolved from the circled zero.

5. a. The 7 is in the hundredths place. Its value is 7/100.

    b. The 6 is in the ten-thousandths place. Its value is 6/10000.

    c. The first 3 is in the thousandths place. Its value is 3/1000.

    d. The first 9 is in the tenths place. Its value is 9/10.

7. a. The decimal for 33/100 is .33. Its name is thirty-three hundredths.

    b. For 392/1000, it is .0392, which is written as three hundred ninety-two ten-thousandths.

    c. The decimal for 54/1000 is .054. Its name is fifty-four thousandths.

    d. The decimal for 7481/10 is 748.1. Its name is seven hundred forty-eight and one tenth.

9. a. The dollar amount for $347.96 can be written as "Three hundred forty-seven and ninety-six hundredths dollars" or as "Three hundred forty-seven dollars and ninety-six cents".

    b. The dollar amount for $23.50 can be written as "Twenty-three and fifty hundredths dollars" or as "Twenty-three dollars and fifty cents".

    c. The amount for $1144.03 can be written "One thousand one hundred forty-four and three hundredths dollars" or "One thousand one hundred forty-four dollars and three cents".

11. Marks have been placed on the line to mark off tenths. The distance between each mark is one tenth of a unit. The location of .07 is between the mark for 0 and the mark for 1/10 (or 0.1). It should be a bit closer to the .1 mark than to the 0 mark. The location of .72 is just to the right of the mark for 7/10. The location of 1.40 is right on the mark that labels 4/10 beyond 1. The location of 1.68 is just to the left of the mark for 1.7 (or 1 and 7/10). Point A is at approximately .65 since it is between the marks for 6/10 and 7/10. It is halfway between .60 and .70. Point B is between 1.5 and 1.6. If it is at the midpoint its value is 1.55. Point C is on the mark for the third tenth after 2. It represents the number 2.3.

13. a. Written in terms of hundredths, .4 is equivalent to .40. A decimal square for .4 shows 4 out of 10 parts shaded. A decimal square for .40 shows 40 out of 100 parts shaded.

   b. Written in terms of thousandths, .47 is equivalent to .470. A decimal square for .47 shows 47 out of 100 parts shaded. A decimal square for .470 shows 470 out of 1000 parts shaded.

15. a. The decimals .7 and .70 are equivalent because .7 can be represented by a decimal square with 7 of 10 columns shaded and .70 is a decimal square with 70 of 100 squares shaded. Shading 70 squares is the same amount as shading 7 columns.

   b. The decimals .43 and .430 are equivalent because .43 can be represented by a decimal square with 43 of 100 squares shaded. To represent .430 on a decimal square we use a square which has been subdivided into thousandths. Each of the small hundredths squares then contains 10 smaller pieces. We shade 430 of these smaller pieces by shading 43 of the small hundredths squares, so .43 and .430 are the same amount.

   c. The decimal square for .45 contains 45 out of 100 pieces shaded. Comparing this to a decimal square for tenths we see that it is more than 4 tenths but less than 5 tenths. It is also less than 6 tenths, so .45 < .6.

17. a. If each small square represents 1 unit, then the longs are 10's and the large squares are 100's. So the figure represents 2 hundreds, 4 tens and 7 ones, or 247.

   b. If each large square represents 1 unit, then the longs are tenths and the small squares are hundredths. In this case the number represented is 2.47.

19. a. To sketch a decimal square to represent 1/4 we need to use a square showing hundredths (or smaller pieces such as thousandths). We could use tenths if we are willing to shade half of one of the tenths. Since 1/4 of 100 is 25, shading 25 of the 100$^{ths}$ squares shows the fraction 1/4. The decimal for 1/4 is .25. The decimal squares below show two of the many different ways to represent 1/4 with decimal squares.

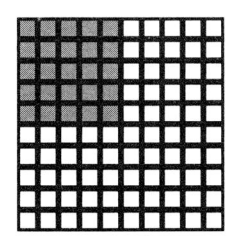

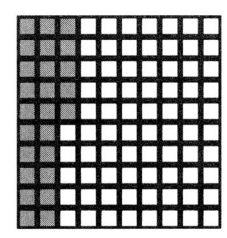

   b. Since 1/5 = 2/10 we can represent 1/5 with a decimal square showing tenths. We can also use hundredths since 1/5 = 20/100, or thousandths since 1/5 = 200/1000. Two different representations of 1/5 = .2 are shown.

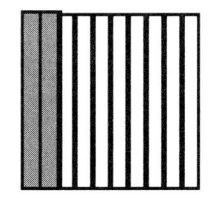

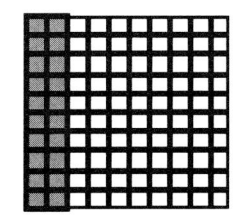

21. a. To find the decimal representation for 3/8 using a calculator, enter 3 ÷ 8. Or divide 3 by 8 using pencil and paper. The result is 0.375.

  b. The decimal for 2/11 has a repeating pattern of digits. Its decimal representation is $0.\overline{18}$

  c. The decimal for 345/990 repeats the pattern 48 infinitely. $345/990 = .3\overline{48}$

  d. The decimal for 15/4 does not repeat. $15/4 = 375/100 = 3.75$.

23. The matches are shown below. They can be found by dividing numerators by denominators or by finding equivalent fractions for the decimals.

$$\frac{1}{3} = .\overline{3} \qquad .25 = \frac{1}{4} \qquad .875 = \frac{7}{8} \qquad .0\overline{6} = \frac{1}{15}$$

$$\frac{5}{6} = .8\overline{3} \qquad .41\overline{6} = \frac{5}{12} \qquad \frac{3}{10} = .3 \qquad \frac{34}{99} = .\overline{34}$$

25. To determine whether a fraction has a terminating decimal, look at the simplest form of the fraction. If the prime factorization of the denominator in this fraction contains only factors of 2 and 5, then the decimal will terminate. Otherwise it will not terminate. [Why?]

  a. The denominator in 7/17 is not a multiple of 2's and 5's. The decimal for 7/17 will be repeating.

  b. Nine has 3 as a factor, so the decimal for 2/9 will be repeating.

  c. The fraction 6/15 is equivalent to 2/5. Since the denominator of 2/5 contains no factors other than 2 or 5, the decimal for 6/15 is terminating.

27. a. To shade the decimal square with 100 parts to represent 1/3 we shade 33 squares. This represents 1/3 of 99 of the squares. We need to shade 1/3 of one of the remaining small squares to exactly represent 1/3. Rounded to the nearest hundredth, $1/3 \approx .33$.

  b. To shade approximately 1/9 we shade 11 parts out of 100. To be exact we would shade 11 squares and one ninth of another square. To the nearest hundredth, $1/9 \approx .11$.

  c. To shade approximately 1/8 we shade 13 parts out of 100. To be exact we would shade 12 squares and one half of another square. To the nearest hundredth, $1/8 \approx .13$.

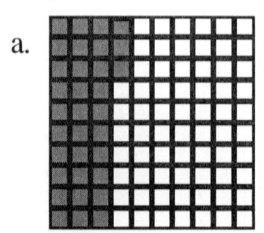

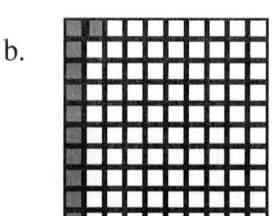

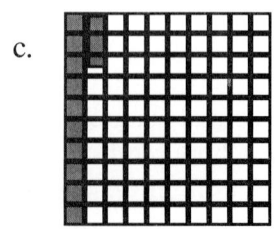

## 6.1 Decimals and Rational Numbers

29. a. For an approximation using <u>only</u> the first nonzero digit we get .0045 ≈ .004.
    If we <u>round</u> to the first nonzero digit we get .005.

    b. For an approximation using <u>only</u> the first nonzero digit we get 0.408 ≈ 0.4.
    If we <u>round</u> to the first nonzero digit we also get 0.4.

    c. For an approximation using <u>only</u> the first nonzero digit we get .074 ≈ .07.
    If we <u>round</u> to the first nonzero digit we also get .07.

    d. For an approximation using <u>only</u> the first nonzero digit we get .00263 ≈ .002.
    If we <u>round</u> to the first nonzero digit we get .003.

31. As decimals, 16/20 = .8, 19/34 ≈ .5588, 38/52 ≈ .73, 21/25 = .84, and 11/17 ≈ .647.

    The decimals in order from smallest to largest are: .5588  .647  .73  .8  .84.

    The fractions arranged in increasing order then are: 19/34, 11/17, 38/52, 16/20, 21/25.

33. a. Divide 1 ÷ 16 to get .0625. This is already to the ten-thousandths place.

    b. Dividing 3 ÷ 32 gives .09375. Rounding to ten-thousandths gives .0938.

    c. Dividing 7 ÷ 64 gives .109375. Rounding to ten-thousandths gives .1094.

    d. Dividing 35 ÷ 64 gives .546875. Rounding to ten-thousandths gives .5469.

35. a. Since the number following the hundredths place is a 2, .3728 rounds to .37.

    b. There is an 8 in the hundredths place, so the 0 in the tenths rounds up. .084 ≈ .1.

    c. Because of the 6 in the ten-thousandths place, 14.3716 ≈ 14.372.

    d. Because we are rounding to tenths, the 9 in the thousandths place has no effect.
    There is a 4 in the hundredths place, so .349 ≈ .3.

37. a. If we write all of these numbers as thousandths we have: .070, 1.003, .080, .075, .300.
    Shaded as decimal squares for thousandths we see that the first one is the smallest because it has only 70 thousandths shaded. The next smallest is .075, with 75 thousandths shaded.

    b. Students may have thought that .075 is smaller than .07 by reasoning that more decimal places means a smaller number. They probably did not have a mental model such as decimal squares available to help them make a judgment.

39. Here are matches of equivalent rational numbers. They can be found by performing the indicated division in the fractions or by writing the decimals as fractions and simplifying.

$.05 = \frac{1}{20}$   $\frac{2}{3} = .\overline{6}$   $.125 = \frac{1}{8}$   $.4 = \frac{2}{5}$

$.58\overline{3} = \frac{7}{12}$   $\frac{7}{18} = .3\overline{8}$   $.\overline{532} = \frac{532}{999}$   $.2\overline{6} = \frac{4}{15}$

41. If your calculator shows at least three more digits you can find the answer there. If not, then long division will also work.

   a. $2 \div 19 \approx 0.1052631579$   b. $14 \div 29 \approx 0.4827586207$

43. a. Rounding to the nearest tenth we get: $17.31 \approx 17.3$   $16.25 \approx 16.3$
   $15.90 \approx 15.9$   $28.06 \approx 28.1$   $22.55 \approx 22.6$

   b. There were more injuries to the knees because $.20 > .025$. The decimal .20 can be represented by a decimal square with 2 tenths shaded. The decimal square for .025 contains less than one tenth.

   c. Half way between .8 and .9 is .85. If the space between the marks for .8 and .9 is marked off into ten equal spaces, the fifth mark will be the halfway point. This represents 8 tenths and 5 hundredths.

45. Here are the decimal equivalents for Fiona's drill bits:
   5/32 = .15625   1/8 = .125   3/16 = .1875   1/4 = .25   1/16 = .0625   3/32 = .09375

   a. The closest drill bit to .13 is .125, the 1/8 inch drill bit.

   b. She does not have a drill bit large enough for size .32.

   c. Her smallest drill bit is 1/16 of an inch.

47. a. There are 12 inches in a foot, so 11 inches is 11/12 of a foot. As a decimal, $11/12 \approx .917$. Lester Steers jumped about 6.92 feet in 1941.

   b. Nine and a half inches is the same as 9.5 inches and $9.5 \div 12 \approx .792$. So Heike Henkel jumped about 6.79 feet in 1982.

49. One pattern in the equations could be described this way: When the denominator is 9, the fraction is equivalent to a repeating decimal that repeats the numerator. This pattern holds for fractions with a one-digit numerator and a denominator of 9. For example, 7/9 = .77777777 . . . and 6/9 = .666666666 . . . It is also true that 9/9 = .999999999 . . . and both of these expressions are equal to 1. But if the numerator is more than 9 the pattern is different. For example, 10/9 = 1.111111 . . . (It is not .1010101010 . . .) And, 11/9 = 1.22222 . . .   *[How would you describe the continuing pattern?]*

## Section 6.2

1. a. All of the temperatures shown on the graph are below zero. They are all represented by negative numbers. The one that is closest to zero is ⁻11.9°, so it is the highest temperature.

   b. The highest temperature is ⁻11.9° and the lowest temperature is ⁻16.3°. Estimating the difference we see that the difference between ⁻12 and ⁻16 is 4. To calculate the actual difference we can subtract: ⁻11.9 – ⁻16.3 = ⁻11.9 + 16.3 = 4.4°.

3. a. A decimal square for .3 contains 3 out of 10 parts shaded. This is equivalent to a hundredths square with 30 parts out of 100 shaded. To add .45, combine the 30 hundredths with 45 hundredths to get a square with 75 out of 100 parts shaded.

   b. A decimal square for .350 consists of 350 out of 1000 parts shaded. This could be shown by looking only at the 10 columns in a tenths square. To show .350 here we would shade three and a half columns. To subtract .2, take away two of the three and a half columns leaving one and a half columns shaded. .350 – .2 = .15   (or .150).

   c. To multiply 3 × .65, take three copies of a decimal square for .65 and combine them together. A .65 decimal square has 65 out of 100 of the small squares shaded. Since 3 × 65 = 195, the three copies together have 195 small squares shaded. This would fill one decimal square and almost fill another one. It shows that 3 × .65 = 1.95.

   d. To show 10 × 3.7 we can think the same way as in part c, looking at 10 copies of three whole squares and 7/10 of a square. This gives 30 + 70/10, or 37. Another way to think about this one is to realize that multiplying by 10 changes units to tens and tenths to units. So instead of 3 units and 7 tenths we have 3 tens and 7 units, or 37.

5. a. Regrouping is needed in the tenths column. 8/10 + 7/10 = 15/10. Since 15/10 is the same as $1\frac{5}{10}$, we regroup 10 of the tenths as one more unit. See the equations below.

$$\begin{array}{r} \overset{1}{4.821} \\ +\,61.73 \\ \hline 66.551 \end{array} \qquad \frac{8}{10}+\frac{7}{10}=\frac{15}{10}=\frac{10}{10}+\frac{5}{10}=1+\frac{5}{10}$$

   b. Regrouping is needed in the thousandths column. 7/1000 + 5/1000 + 9/1000 = 21/1000. Since 21/1000 is the same as 2/100 + 1/1000, we regroup 20 of the thousandths as two more hundredths. See the equations below.

$$\begin{array}{r} .3\overset{2}{6}7 \\ .015 \\ +\,.509 \\ \hline .891 \end{array} \qquad \frac{7}{1000}+\frac{5}{1000}+\frac{9}{1000}=\frac{21}{1000}=\frac{20}{1000}+\frac{1}{1000} \\ =\frac{2}{100}+\frac{1}{1000}$$

7. To illustrate the products we use an area model. The dimensions of the rectangle are the two factors being multiplied. The area of the rectangle is the product. In the diagram here for 1.7 × 2.2 one of the dimensions is 1 unit and 7 tenths and the other is 2 units and 2 tenths. The area contains 2 whole units of area, two sections each of .7, one section of .2 and one section with area .14. Together this adds to an area of 3.74. Create a similar model for part b.

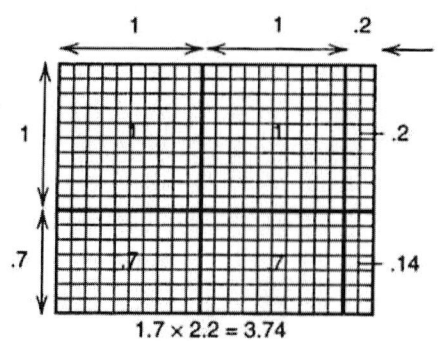

9. a. To multiply 3.2 × 7.8, first multiply 32 × 78, then since there are a total of two digits behind the decimal points in the factors, we leave two digits behind the decimal point in the product. Since 32 × 78 = 2496, we know that 3.2 × 7.8 = 24.96.

   b. Because of the properties of equal quotients, dividing 1.4146 ÷ .22 is the same as dividing 141.46 ÷ 22. We have multiplied both dividend and divisor by 100. Since 141.46 ÷ 22 = 6.43, then 1.4146 ÷ .22 = 6.43.

11. a. To calculate .01 × 7.6 mentally we may recognize that multiplying by .01 is the same as dividing by 100 (because we are multiplying by 1/100). Since we are dividing by the second power of 10, we move the decimal two places to the left. The answer is .076. Hint: If you are unsure on which way to move the decimal, determine whether you are multiplying by a number that is more than 1 or less than 1. In this case we multiplied by .01, which is less than 1. So our answer needed to be a number that is less than 7.6.

    b. To multiply by .001 move the decimal move three places to the left. .001 × 34 = .034.

    c. To divide by 100 move the decimal point two places to the left. .03 ÷ 100 = .0003.

    d. To divide by 10 move the decimal point one place to the left. .04 ÷ 10 = .004.

13. To convert repeating decimals to fraction form, we use the fact that .1111 ... = 1/9, .01010101 ... = 1/99, .001001001 ... = 1/999, etc.

    a. .5555 ... = 5 × .1111 ... = 5 × 1/9 = 5/9.

    b. .141414 ... = 14 × .010101 ... = 14 × 1/99 = 14/99.

    c. .1555 ... = 1/10 × 1.5555 ... = 1/10 × (1 + .5555 ...) = 1/10 + 1/10 × 5/9
       = 1/10 + 5/90 = 14/90.

15. a. For a decimal to be between .6 and .7, its fraction value must be between 6/10 and 7/10, or equivalently, between 60/100 and 70/100. Any decimal that begins .6-- will work. Some examples of correct answers are .61, .65, .632, .60001, etc.

    b. Examples of numbers between .005 and .006 include: .0052, .0055, .0059, .005987.

17. a. One way to calculate .337 − .294 mentally is to use an add up method. We add .006 to .294 to get to .3. Then add .037 to get to .337. The difference is .006 + .037 = .043.

   b. Since 4.3 + .7 = 5, we are just adding an additional tenth: 4.3 + .8 = 5.1.

   c. Multiplying by 101 is the same as multiplying by 100 and by 1, and adding the two products. $2.6 \times 101 = 2.6 \times 100 + 2.6 \times 1 = 260 + 2.6 = 262.6$.

   d. Using equal quotients, $12.9 \div 300 = .129 \div 3 = .043$. Or, instead of dividing both numbers by 100 we can divide both by 3 to get $4.3 \div 100 = .043$.

19. a. $.25 \times 48 = \frac{1}{4} \times 48 = 12$   b. $.5 \times 40.8 = \frac{1}{2} \times 40.8 = 20.4$   c. $5.5 \times .2 = 5.5 \times \frac{1}{5} = 1.1$

21. a. Using front-end estimation we add $20 + $40 + $20 + $10 = $90.
      Rounding to the leading digit gives us $30 + $50 + $20 + $10 = $110.

   b. Using front-end estimation we subtract $400 − $100 = $300.
      Rounding to the leading digit gives us $500 − $100 = $400.

23. a. We estimate $3.1 \times 4.9$ by rounding 3.1 to 3 and 4.9 to 5 to get $3.1 \times 4.9 \approx 3 \times 5 = 15$. From the diagram below we can see that the amount of area lost by rounding 3.1 to 3 is a little bit more than the amount of area gained by rounding 4.9 to 5. So by rounding we have decreased the total area of the rectangle. Our estimate will be low. The actual product is a little more than 15.

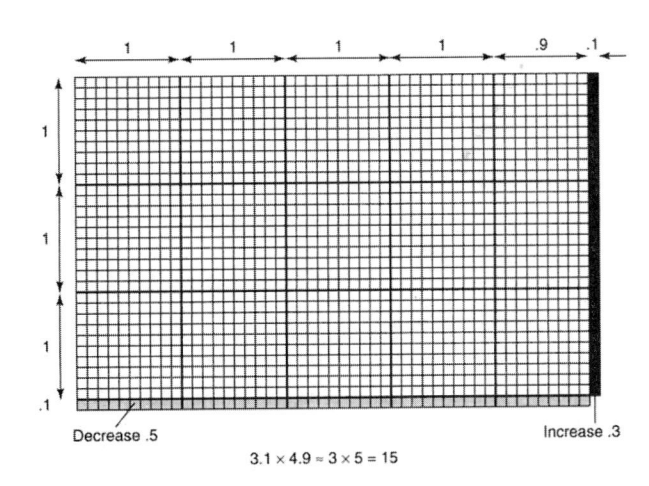

   b. $5.3 \times 1.6 \approx 5 \times 2 = 10$. The diagram at right shows that by rounding 1.6 to 2 we increase the area of the rectangle more than we decrease it by rounding 5.3 to 5. The approximate answer of 10 will be too high.

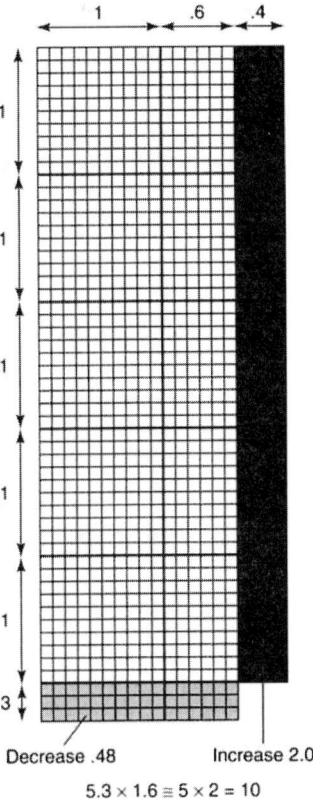

25. a. To estimate $8 \div .48$ we round .48 to .5, which is the same as 1/2.
So $8 \div .48 \approx 8 \div 1/2 = 8 \times 2 = 16$.

   b. $11.63 + .4 \approx 11.6 + .4 = 12$.

   c. Since .34 is close to .3333 . . . we can approximate .34 with 1/3. Mentally compute $1/3 \times 120$ as one third of 120. So $.34 \times 120 \approx 1/3 \times 120 = 40$.

   d. Round .23 to .25, or 1/4 and round 81.6 to 80.
Then $.23 \times 81.6 \approx 1/4 \times 80 = 20$.

27. a. Using either rounding or front-end estimation, we get $3.04 \times 5.3 \approx 3 \times 5 = 15$. Of the choices given, 16 is the only reasonable estimate. The total percent of those selecting the other responses was $28 + 18 + 23 + 9 = 78\%$. Note that the percentages given add to 99 instead of 100. This is probably due to rounding. With this in mind, 79% is probably a better answer to this question.

   b. Some students may have estimated the answer by thinking of $300 \times 50$. If so, they probably do not understand that the symbols 3.04 indicate a number close to 3.

29. a. This student may have learned to bring the decimal point straight down when adding or subtracting. This is not incorrect, but we need to know more than this. We need to know that when we are adding 4 tenths and 8 tenths we will have 12 tenths, not 12 hundredths.

   b. This is a very common error. The student sees no number to subtract the 6 from and follows the path of least resistance by just bringing the 6 down.

   c. This student misplaced the decimal point in the product. They may have been thinking again that the decimal should go straight down.

   d. The student was not careful with putting the digits in the quotient in the proper place value. A zero is needed between the 6 and the 2. This kind of mistake can also be made with whole number division. Approximating the answer is one way to avoid this mistake.

31. a. Adding .0007 to 273.5186 gives 273.5193.

   b. Subtracting .0006 from 273.5186 gives 273.5180

   c. Subtracting .503 from 273.5186 gives 273.0156.

33. a. Using a calculator, we divide $1 \div 2.4$ and get .416666 . . . Rounded to five places this is .41667. Multiplying $2.4 \times .41667 = 1.000008$. It is not exactly 1 due to the rounding.

   b. $1 \div .48 = 2.083333$ . . . Rounding, we get 2.08333. And $2.08333 \times .48 = .9999984$.

   c. $1 \div .0046 \approx 217.39130$. Multiplying, $217.39130 \times .0046 = .99999998$.

35. a. In screen 4 we get the answer to $^-22.496 \times {}^-1.6$. This is 35.9936
In screen 5 we get the answer to $35.9936 \times {}^-1.6$. This is $^-57.58976$

   b. In screen 4 we get the answer to $456.5 - 16.3$. This is 440.2
In screen 5 we get the answer to $440.2 - 16.3$. This is 423.9

## 6.2 Operations with Decimals

37. a. We can immediately see that this sequence is not arithmetic, because the difference between consecutive numbers is not remaining constant. We check to see if there is a common ratio by dividing consecutive numbers. $7.56 \div 4.2 = 1.8$. Also, $7.56 \times 1.8 = 13.608$ and $13.608 \times 1.8 = 24.4944$. So the sequence is geometric and has common ratio 1.8. The next number in the sequence is $24.4944 \times 1.8 = 44.08992$.

   b. This sequence is arithmetic. There is a common difference of $^-0.7$ because each number is obtained by subtracting 0.7 from the previous number. The next number is 13.5.

39. For FICA/OASI Michelle pays $61150 \times .062 = \$3791.30$. For FICA/Medicare she pays $61150 \times .0145 = 886.675 \approx \$886.68$.
    Her total FICA taxes are $\$3791.30 + \$886.68 = \$4677.98$.

41. a. There are three different rates that apply to the finance charge on $1200. The first $500 costs $500 \times .0125 = \$6.25$. The second $500 costs $500 \times .0095 = \$4.75$. The last $200 costs $200 \times .0083 = \$1.66$. The total finance charge is $\$6.25 + \$4.75 + \$1.66 = \$12.66$

   b. If $75.25 is the amount past due, then the late charges are $75.25 \times .05 = 3.7625$. Rounding to the nearest cent we get $3.76.

43. a. A decrease in consumption is indicated by a negative number and an increase by a positive number. Look for a year in which the numbers for coal and petroleum are negative, but the number for natural gas is positive. This was 1991.

   b. There were increases in the demand for coal in the seven consecutive years from 1992 to 1998. The total of these numbers giving the percent changes for each year is $.5 + 2.9 + .6 + .4 + 6.8 + 2.9 + .9 = 15.0$. *[Does this mean that the demand for coal increased by exactly 15% during this time period?]*

   c. Consumption of natural gas decreased the most in 1998. During that year, consumption of coal increased by 0.9% and consumption of petroleum increased by 0.5%.

45. a. There are two ways to solve this problem. One way is to first compute the monthly costs of electricity for the two types of televisions and then find the difference. The b/w TV costs $29.6 \times .04 = \$1.184$ per month. The color TV costs $55.0 \times .04 = \$2.20$ per month. The difference is $2.20 - 1.184 = \$1.016 \approx \$1.02$.
   Another way to solve the problem is to find the difference in Kilowatt-hours per month first. This is $55.0 - 29.6 = 25.4$. So the difference in cost is $25.4 \times .04 = 1.016 \approx \$1.02$.

   b. The savings in kilowatt-hours per month is $152.4 - 94.7 = 57.7$. In one year this amounts to a savings of $57.7 \times 12 = 692.4$ kilowatt-hours.
   In dollars saved this is $692.4 \times .04 = 27.696 \approx \$27.70$.
   A slightly different amount will be obtained if we first compute the savings in dollars per month and round to the nearest cent before multiplying by 12.

   c. The savings in kilowatt-hours per month is $97.6 - 15.8 = 81.8$. In one year this amounts to a savings of $81.8 \times 12 = 981.6$ kilowatt-hours.
   In dollars saved this is $981.6 \times .04 = 39.264 \approx \$39.26$.
   If the monthly dollar amount is found first and rounded, then the yearly total is $39.24.

47. a. Janet Evans' time was 4:03.85. Petra Thurner's time was 4:09.89. The difference in these times was approximately 6 seconds. It was exactly 9.89 – 3.85 = 6.04 seconds.

   b. The difference between 4:07.18 and 4:03.85 is 7.18 – 3.85 = 3.33 seconds.

49. a. In terms of cents, half of one cent is .5 cent. (In terms of dollars it is $0.005).

   b. We can multiply .05 by 650,000 to find the number of cents higher for the bid. 650,000 × .05 = 32,500 cents. In dollars this is $325.00.

   c. If the bid was .5 of a cent higher, then it was 650,000 × .5 = 325,000 cents higher. In dollars this is $3,250.00.

   d. By misplacing the decimal point the school district paid $3250 extra rather than the $325 extra that they thought they were paying. They lost 3250 – 325 = $2925. (One could also argue that they lost $3250 because they would have gone with the lower bid if they hadn't misplaced the decimal point.)

## Section 6.3

1. a. There are 4 empty seats for every 1 full seat. So 4 out of 5 seats are empty. The fraction is 4/5.

   b. One-fifth of the seats are filled. One-fifth of 9,000,000 is 1,800,000. This can be found in any of these ways: 9,000,000 ÷ 5;  1/5 × 9,000,000;  9,000,000 × 0.2.

3. a. This problem can be solved by using a proportion. $\frac{11}{3} = \frac{407}{x}$

   Solving, we get 11x = 407(3). So 11x = 1221. Then x = 111. There were 111 patent applications received by non citizens.

   b. Erik types twice as fast as Trenton, so Erik types 50 words every minute. In 3 minutes, Erik will type 150 words.

5. a. One way to solve this problem is to reason as follows. If 1.5 pounds of fish costs $3.12 then .5 pound costs 1/3 of this, or $1.04. So 3.5 pounds costs $1.04 × 7 = $7.28. Another approach is to write and solve a proportion.
   $\frac{1.5}{3.12} = \frac{3.5}{x}$     So 1.5 x = 3.12 × 3.5. Hence 1.5 x = 10.92, so x = $7.28.

   b. Using the proportion $\frac{8}{2.66} = \frac{20}{x}$ we get 8x = 2.66 × 20.    So x = $6.65. Another approach is to reason that if 8 ounces cost $2.66, then 4 ounces cost $1.33. So 20 ounces cost five times as much. A diagram may also be helpful.

   c. If 10 pounds of nails cost $4.38 then 1 pound of nails costs $.438. Multiplying the cost for 1 pound by 3.2 we get the cost for 3.2 pounds. $.438 × 3.2 = $1.4016 ≈ $1.40.

7.  a. $37\frac{1}{2}\% = .375$  A 10 by 10 decimal square to represent $37\frac{1}{2}\%$ would have $37\frac{1}{2}$ of the 100 squares shaded.

    b. $6.5\% = .065$  A 10 by 10 decimal square to represent 6.5 % would have $6\frac{1}{2}$ of the 100 squares shaded.

    c. $28\frac{1}{3}\% = .28333\ldots$  A 10 by 10 decimal square to represent $28\frac{1}{3}\%$ would have $28\frac{1}{3}$ of the 100 squares shaded.

9.  a. A decimal square for .60 has 60 out of 100 squares shaded.  .60 = 60%.

    b. A decimal square for .06 has 6 out of 100 squares shaded.  .06 = 6%.

    c. A decimal square for .256 has 256 out of 1000 parts shaded.  If it is a square containing 100 small squares it has 25.6 small squares shaded.  .256 = 25.6%.

11. a. 4/5 = 80/100, so 4/5 = 80%.  We can also calculate 4 ÷ 5 to find that 4/5 = .8 = 80%.

    b. Using a calculator, 5 ÷ 6 ≈ .83333.  So 5/6 ≈ 83.3%.

    c. 7/4 = 1.75.  This is 175/100, so 7/4 = 175%.

13. a.

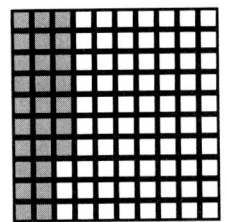

    27 out of 100 small squares are shaded to represent 27%.

    In this figure the whole square represents 160.  So each small square represents 160 ÷ 100 = 1.6.  Since 27 small squares are shaded the shaded area represents 27 × 1.6 = 43.2.  So 27% of 160 = 43.2.

    b.

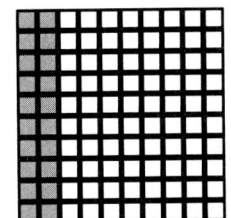

    20 small squares shaded represent 20%.

    If the whole square represents 200, then each small square represents 2.
    20 small squares represent 40, so 20% of 200 is 40.

c. If 10 small squares represent 4, then 100 small squares represent 40. If 10% of a number is 4, then the number is 40.

d. 100% of 65 is 65. 40% of 65 is equal to 65 × .4 = 26. So 140% of 65 is 65 + 26 = 91.

e. Since 75 is more than 50, it is more than 100% of 50. If one whole decimal square represents 50, then $1\frac{1}{2}$ decimal squares represents 75. 75 is 150% of 50.

15. a. Since 10% is 1/10 we can find 10% (or 1/10) of a decimal number by shifting the decimal point one place to the left. 10% of $128.50 = $12.85.

  b. 75% is 3/4, so 75% of 32 = 3/4 of 32 = 3 × 1/4 of 32 = 3 × 8 = 24.

  c. 10% of $60 is $6, so 90% of $60 is $6 less than $60. 90% of 60 = 60 − 6 = $54.

  d. 10% of 80 is 8. So 110% of 80 is 80 + 8 = 88.

17. a. 9% of $30.75 ≈ 10% of $31 = $3.10.

  b. 19% of 60 ≈ 20% of 60 = 1/5 of 60 (or, .2 × 60) = 12.

  c. 4.9% of 128 ≈ 5% of 130 = 1/2 of 10% of 130 = 1/2 of 13 = 6.5.

  d. 15% of 241 ≈ 15% of 240 = 10% of 240 plus 1/2 of 10% of 240 = 24 + 12 = 36.

19. a. 2/19 ≈ 2/20 = 1/10 = 10%.

  b. 408/1210 ≈ 400/1200 = 1/3 ≈ 33%.

  c. 100/982 ≈ 100/1000 = 1/10 = 10%.

21. To convert to scientific notation, we place a decimal point after the first nonzero digit to create a mantissa between 1 and 10. Then count the number of places the decimal point has moved in relation to the original number. This gives the characteristic (power of 10).

  a. $4{,}600{,}000{,}000 = 4.6 \times 10^9$    b. $.000000027 = 2.7 \times 10^{-8}$

23. a. $1.2 \times 10^{-9} = .0000000012$ (Note that there are 8 zeros before the one, but the decimal point has moved 9 places to the left.)

  b. $3.15569 \times 10^7 = 31{,}556{,}900$ (Because the power of 10 is 7, we know that this number is between 10 million and 100 million.)

25. a. Because multiplication is commutative and associative,
$5.868 \times 10^{13} \times 2.7 \times 10^4 = (5.868 \times 2.7) \times (10^{13} \times 10^4) = 15.8436 \times 10^{17}$.
This last number is not quite in scientific notation. It is equivalent to $1.58436 \times 10^{18}$.
This is the distance in miles from the sun to the center of our galaxy.

b. To compute $(4.619 \times 10^8) \div (3.1 \times 10^5)$ we can divide the mantissas and characteristics separately. $4.619 \div 3.1 = 1.49$. To divide $10^8 \div 10^5$ we can subtract the exponents. $10^8 \div 10^5 = 10^3$. So $(4.619 \times 10^8) \div (3.1 \times 10^5) = 1.49 \times 10^3$. It took 1490 seconds for the signals to reach the earth. (about 25 min.)

27. a. 4.3% of the 1180 injured football players had lower back injuries. We can find 4.3% of 1180 by multiplying $1180 \times .043$. The result is 50.74, so about 51 players had lower back injuries.

b. 6.0% of $1180 = 1180 \times .06 = 70.8$. Approximately 71 players had injuries to the feet and toes.

29. Since there are more students than teachers, the student to teacher ratio will be more than 1. To calculate this ratio we divide number of students by number of teachers.

a. For Maine it is $206{,}000 \div 16{,}000 \approx 12.9$.
For Missouri it is $910{,}000 \div 66{,}200 \approx 13.7$.
For Oregon it is $551{,}000 \div 27{,}100 \approx 20.3$.
For Wyoming it is $88{,}000 \div 6{,}500 \approx 13.5$.

b. Of these four states the one with the best student to teacher ratio is Maine.

c. Of these four states the one with the poorest student to teacher ratio is Oregon.

31. a. A 70% discount means that the sale price is 30% of the original price. The diagram below shows that the price before the discount was very close to $60, because if 30% of the price is about $18, then 10% of the price is about $6.

```
        30%              70%
      ┌──┬──┬──┬──┬──┬──┬──┬──┬──┬──┐
      │6 │6 │6 │6 │6 │6 │6 │6 │6 │6 │
      └──┴──┴──┴──┴──┴──┴──┴──┴──┴──┘
      $17.99
```

We can also find the exact price before the discount using an equation. If p represents the original price, then 30% of p is $17.99, or, $.3 \times p = 17.99$.
Then $p = 17.99 \div .3 \approx \$59.97$.

b. The amount of the discount was $99.99 - \$79.99 = \$20$. To find the percent of discount we need to find what percent $20 is of the <u>original</u> price $99.99.
$20 \div 99.99 = .200200200\ldots \approx .200 = 20.0\%$.
(Since the original price was almost exactly $100, we could have said that the percent discount was 20% as soon as we knew that the amount of discount was $20.)

c. We need to find 107.5% of $32,000. We multiply $32,000 × 1.075 = $34,400.

d. If 6 of 28 did not enroll, then 22 of 28 did enroll. We need to find what percent 22 is of 28. We divide 22 ÷ 28 ≈ .7857 ≈ 78.6%.

33. a. To find the cost in cents per ounce we divide the number of cents in the price by the weight in ounces. For the small package, 119 ÷ 32 ≈ 3.72. Each ounce costs about 3.72¢.

   b. Since the cost per ounce for the large package is 1.85 ¢, the large package is the better buy.

   c. The ratio of amount of flour in the small package to the large package is 32 to 80, which is equivalent to 2 to 5. In other words, the small package will make 2/5 as much as the large one will. Since 2/5 of 10 is 4, the small package is enough flour for 4 batches.

35. The first equation uses the multiplicative identity by rewriting 9.85 as 1 × 9.85. In the second equation the distributive property of multiplication over subtraction is used.

   a. 10% of $209.50 is $20.95. 5% is half of this, so 5% of $209.50 is $20.95 ÷ 2 ≈ $10.48. Then 15% of $209.50 is $20.95 + $10.48 = $31.43. So we deduct $31.43 from $209.50 to get the sale price of $178.07.
   Or, alternatively, calculate 85% of $209.50. $209.50 × .85 ≈ $178.08.

   b. The price after a 20% discount is 80% of the original price. $153.95 × .8 = $123.16.

   c. The price after a 28% discount is 72% of the original price. $86 × .72 = $61.92.

37. a. After one year with 4% inflation, the washing machine will cost $545 × 1.04 = $566.80. The price increased by 566.80 – 545.00 = $21.80. ($21.80 is 4% of $545.)

   b. After five years of 4% inflation, the cost of the machine will be
   545 × 1.04 × 1.04 × 1.04 × 1.04 × 1.04 ≈ $663.08. The price increased by 663.08 – 545 = $118.08. *[Note that this is more than an increase of 5 x 4 = 20%. Why does it increase more than 20% even though it is caused by 5 years of 4% inflation?]*

   c. One way to solve this is with a calculator by starting with 545 and repeatedly multiplying by 1.04 until surpassing 745. Counting the number of factors of 1.04 we find that after eight years we get to about $746. If your calculator has a power key you can also use exponents to find the answer. Guessing and checking several powers of 1.04 we eventually find that 545 × $1.04^8$ ≈ 745.87.

   d. $28,500 × 1.05 × 1.05 × 1.05 ≈ $32,992.31.

39. a. The interest for one year is 5% of $274. $274 × .05 = $13.70 (or, 60 × .05 = 3 pounds).

   b. The total amount owed at the end of the first year was $274 + $13.70 = $287.70. 10% of $287.70 is $28.77.

   c. To find the value of 1000 pounds at 5% interest after five years, calculate 1000 × 1.05 × 1.05 × 1.05 × 1.05 × 1.05 ≈ $1276.28.  Or, calculate $1000 \times 1.05^5$.
   *[If your calculator can compute $1000 \times 1.05^5$, check Franklin's statement.]*

41. Here are the missing amounts from the chart. Astronomical units are found by dividing the distance in miles by 93,003,000 (or $9.3003 \times 10^7$). For example, to find the distance in astronomical units from the sun to Jupiter divide 483,881,000 ÷ 93,003,000.

    | Planet | Scientific Notation | Positional num. | Astronomical Units |
    |---|---|---|---|
    | Mercury: |  | 36,002,000 | .4 unit |
    | Venus: | $6.7273 \times 10^7$ |  | .7 unit |
    | Earth: |  | 93,003,000 | 1.0 unit |
    | Mars: | $1.41709 \times 10^8$ |  | 1.5 units |
    | Jupiter: |  | 483,881,000 | 5.2 units |
    | Saturn: | $8.87151 \times 10^8$ |  | 9.5 units |
    | Uranus: |  | 1,784,838,000 | 19.2 units |
    | Neptune: | $2,796693 \times 10^9$ |  | 30.1 units |
    | Pluto: |  | 3,669,669,000 | 39.5 units |

43. Other than the first and second terms, numbers in the first sequence are found by multiplying the previous number by 2. The first sequence is   0   3   6   12   24   48   96.
    Now add 4 and divide by 10. This sequence is   .4   .7   1.0   1.6   2.8   5.2   10.0
    For example, (3+4) ÷ 10 = .7  and  (48+4) ÷ 10 = 5.2.

    a. Mars is the 4th planet, so Bode's Law predicts a distance of 1.6 astronomical units from the Sun to Mars. It predicts that the next planet would be 2.8 astronomical units for the Sun. This is the approximate location of the asteroid belt.

    b. Extending the sequences above one more place, we get 96 × 2 = 192 in the top sequence. Then we get 192 + 4 = 196 ÷ 10 = 19.6 a.u. for the distance from the Sun to Uranus as predicted by Bode's Law.

## Section 6.4

1. a. $\sqrt{49}$ is rational because it is equal to 7.

   b. Even though there is a pattern in .131131113 . . . , it is not a pattern in which there is a single same block of digits that is repeated. So this number is irrational.

   c. .113113113 . . . is rational because the block of digits 113 is continually repeated. (It is equal to 113/999.)

   d. $\sqrt{14}$ is irrational. Its decimal representation is infinite and does not have a repeating pattern of a single block of digits.

3. a. If we call the length of the missing side c, then since c is the hypotenuse, by the Pythagorean Theorem $\sqrt{8}^2 + 5^2 = c^2$. So $8 + 25 = c^2$ and $c^2 = 33$. That means that $c = \sqrt{33}$.

   b. Again call the hypotenuse c. Then $7^2 + 8^2 = c^2$. So $49 + 64 = c^2$ and $c^2 = 113$. If $c^2 = 113$ then $c = \sqrt{113}$.

5. a. $\sqrt{\frac{1}{16}} = \frac{1}{4}$ because $\frac{1}{4} \times \frac{1}{4} = \frac{1}{16}$     b. $\sqrt[3]{64} = 4$ because $4^3 = 64$.

   c. $\sqrt{9.61} = 3.1$ because $3.1 \times 3.1 = 9.61$     d. $\sqrt[3]{-125} = -5$ because $(-5)^3 = -125$.

7. a. $\sqrt{18}$ is irrational. Using a calculator with a square root key or by a guess and check method we find that $\sqrt{18} \approx 4.2$.

   b. The cube root of 216 is exactly 6 because $6^3 = 216$.

   c. $\sqrt{\frac{1}{9}} = \frac{1}{3}$ because $\frac{1}{3} \times \frac{1}{3} = \frac{1}{9}$

9. a. $\sqrt{7} \approx 2.6$     b. The cube root of $30 \approx 3.1$     c. $\sqrt{3} \approx 1.7$

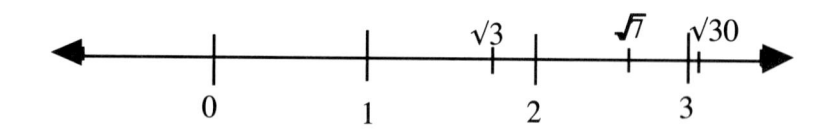

   ```
   ←——+————————+———+——+——+——→
      0         1   √3  2 √7  3 √30
   ```

11. a. To find the Pythagorean triple we substitute the given numbers in place of u and v in each expression to find a, b, and c.
    If $u = 2$ and $v = 1$, then    $a = 2uv = 2 \times 2 \times 1 = 4$
    $b = u^2 - v^2 = 2^2 - 1^2 = 4 - 1 = 3$
    $c = u^2 + v^2 = 2^2 + 1^2 = 4 + 1 = 5$
    Checking $a^2 + b^2 = c^2$, we get $4^2 + 3^2 = 5^2$ or $16 + 9 = 25$. It checks.

    b. If $u = 3$ and $v = 2$, then    $a = 2uv = 2 \times 3 \times 2 = 12$
    $b = u^2 - v^2 = 3^2 - 2^2 = 9 - 4 = 5$
    $c = u^2 + v^2 = 3^2 + 2^2 = 9 + 4 = 13$
    Checking $a^2 + b^2 = c^2$, we get $12^2 + 5^2 = 13^2$ or $144 + 25 = 169$. It checks.

    c. If $u = 6$ and $v = 5$, then    $a = 2uv = 2 \times 6 \times 5 = 60$
    $b = u^2 - v^2 = 6^2 - 5^2 = 36 - 25 = 11$
    $c = u^2 + v^2 = 6^2 + 5^2 = 36 + 25 = 61$
    Checking $a^2 + b^2 = c^2$, we get $60^2 + 11^2 = 61^2$ or $3600 + 121 = 3721$. It checks.

## 6.4 Irrational and Real Numbers

13.

| | Whole Numbers | Integers | Rational Numbers | Real Numbers |
|---|---|---|---|---|
| −3 | | x | x | x |
| 1/8 | | | x | x |
| √3 | | | | x |
| π | | | | x |
| 14 | x | x | x | x |
| 1.6/4 | | | x | x |
| .828282... | | | x | x |

Note that all whole numbers are integers, all integers are rational, and all rational numbers are real numbers. 1.6/4 is rational because it is equal to .4 or 4/10. .828282... is rational because it is equal to 82/99. Neither π nor √3 are rational because neither can be written exactly as a fraction of an integer over an integer.

15. a. Subtraction is not closed on the set of whole numbers. For example, 5 − 7 = ⁻2 and ⁻2 is not a whole number, while 5 and 7 are both whole numbers.

   b. If we divide a rational number by a nonzero rational number we get a rational number. So division by nonzero rational numbers is closed on the set of rational numbers.

17. a. This is an example of the commutative property of multiplication because we are saying that it does not matter in what order we multiply two numbers.

   b. This is an example of the distributive property of multiplication over addition. The multiplication by $\sqrt{7}$ is being distributed to both parts of the sum of 3 and $\sqrt{2}$. In this case the general form is (a+b)c = ac + bc.

   c. This is the associative property of addition. When three numbers are being added we get the same result whether we add the first two numbers first or add the last two numbers first.

19. a. $\sqrt{2} \times \sqrt{20} = \sqrt{40}$. This number is irrational.

   b. $10 + \sqrt{8}$ is irrational. The sum of any rational number plus any irrational number is always irrational.

   c. $\sqrt{6} \times \sqrt{24} = \sqrt{144} = 12$. So this product of two irrational numbers is rational.

21. Look for square factors of the numbers under the radical. For example, 9 is a factor of 45 and 9 is the square of 3.

   a. $\sqrt{45} = \sqrt{9 \times 5} = \sqrt{9} \times \sqrt{5} = 3\sqrt{5}$    b. $\sqrt{48} = \sqrt{16 \times 3} = \sqrt{16} \times \sqrt{3} = 4\sqrt{3}$

   c. $\sqrt{60} = \sqrt{4 \times 15} = \sqrt{4} \times \sqrt{15} = 2\sqrt{15}$

130    *Chapter 6  Decimals: Rational and Irrational Numbers*

23. a. True. This is the theorem given on page 287 and it is the property that we used in #21 and #22.

    b. False. For example, $\sqrt{9} - \sqrt{4} = 3 - 2 = 1$, but $\sqrt{9-4} = \sqrt{5} \neq 1$.

25. a. We can multiply the fraction by any form of 1 without changing its value. In this case we want to multiply by $\sqrt{7}/\sqrt{7}$ because that will make the denominator a rational number.

$$\frac{4}{\sqrt{7}} \times \frac{\sqrt{7}}{\sqrt{7}} = \frac{4\sqrt{7}}{\sqrt{7}\sqrt{7}} = \frac{4\sqrt{7}}{7}$$

    b. Here we multiply by $\frac{\sqrt{6}}{\sqrt{6}}$ to rationalize $2\sqrt{6}$.

$$\frac{3}{2\sqrt{6}} \times \frac{\sqrt{6}}{\sqrt{6}} = \frac{3\sqrt{6}}{2\sqrt{6}\sqrt{6}} = \frac{3\sqrt{6}}{12} = \frac{\sqrt{6}}{4}$$

27. The number that this calculator will display for $\sqrt{1.1892070}$ is 1.0905077. If your calculator has room for more digits in the display your answer may have more accuracy. Continuing to press the square root key will keep giving results that get closer to 1. Eventually the calculator will display the number 1 because it rounds the number to 1.

29. For numbers between zero and one the square root of the number is greater than the number, but less than one. For example, $\sqrt{\frac{1}{4}} = \frac{1}{2}$ and $\sqrt{\frac{9}{16}} = \frac{3}{4}$. By repeatedly taking square roots, eventually the number gets so close to 1 that the calculator rounds it to 1.

31. a. Since $.25 = 1/4$, this is the fourth root. The fourth root of 81 is 3 because $3^4 = 81$.

    b. Since $1 \div 2 = 1/2$, this is the square root. $\sqrt{2.25} = 1.5$
    c. Since $.5 = 1/2$, this is the square root. $\sqrt{3.0625} = 1.75$

    d. Since $1 \div 3 = 1/3$, this is the cube root. The cube root of $^-4.096$ is $^-1.6$.
    (Note: Some calculators will give an error message for this set of keystrokes because they are not programmed to recognize the 1/3 power of a negative number as being a real number. In this case just enter the $^-4.096$ as 4.096 and take the opposite of the answer.)

33. Use the Pythagorean theorem. If x is the number of steps across the field diagonally, then $300^2 + 500^2 = x^2$. So $x^2 = 90,000 + 250,000 = 340,000$. Then $x = \sqrt{340,000} \approx 583$. Ed can save about $800 - 583 = 217$ steps by walking from corner to corner.

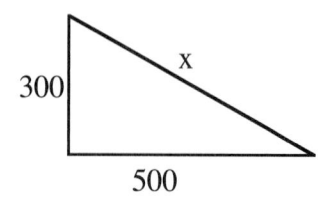

35. The Pythagorean theorem applies because the pipe, the brace, and the ground form a right triangle. If we call the length of the brace x, then $8^2 + 10^2 = x^2$. So $x^2 = 64 + 100 = 164$. Then $x = \sqrt{164} \approx 13$. To the nearest foot the brace should be 13 feet long.

37. The diagram below shows that one of the pieces cut out forms a right triangle with hypotenuse 12 inches long. If we call the lengths of the two legs of the right triangle each x, then by the Pythagorean theorem we have $x^2 + x^2 = 12^2$. Since $x^2 + x^2 = 2x^2$ this is equivalent to the equation $2x^2 = 12^2$. So $2x^2 = 144$ and $x^2 = 72$. Then $x = \sqrt{72} \approx 8.5$ inches. The original square has dimensions 2x by 2x, so it is about 17.0 by 17.0 inches.

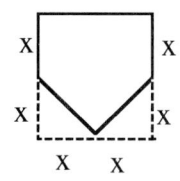

39. a. In the formula that gives the time for one orbit as a function of the apogee and perigee we replace A with 4300 and P with 4100. (A calculator is helpful.)
$$T = \frac{(4300+4100)\sqrt{(4300+4100)}}{501,186} = \frac{8400\sqrt{8400}}{501,186} \approx 1.5 \text{ hours.}$$

b. Replacing both A and P in the formula with 26,300 we get:
$$T = \frac{(26300+26300)\sqrt{(26300+26300)}}{501,186} = \frac{52600\sqrt{52600}}{501,186} \approx 24 \text{ hours.}$$
Since the earth also makes one rotation in a 24 hour period, the satellite will stay in the same position relative to the earth if its orbit takes 24 hours.

41. a. The ratio of the first two Fibonacci numbers is 1/1 = 1.
The ratio of the 3$^{rd}$ number to the 2$^{nd}$ number is 2/1 = 2.
The ratio of the 4$^{th}$ number to the 3$^{rd}$ number is 3/2 = 1.5
Continuing the pattern we get: 5/3 = 1.6666...;   8/5 = 1.6;   13/8 = 1.625;
21/13 ≈ 1.61538;   34/21 ≈ 1.61905;   55/34 ≈ 1.61765

b. A number in the Fibonacci sequence is found by adding the two previous numbers. The sequence continues 1,1,2,3,5,8,13,21,34,55,89,144,233,377,610, . . .
Continuing the pattern of ratios we get:  89/55 ≈ 1.61818; 144/89 ≈ 1.61798;
233/144 ≈ 1.61806;  377/233 ≈ 1.61803;  etc.
The first ratio which is equal to the golden ratio of 1.618033 . . . to the first 4 decimal places is 233/144. (The next one, 377/233, is correct to 5 places.)

## Chapter 6 Test

1. a. A decimal square for .4 has 4 of the 10 parts shaded. This is also equivalent to a square with 40 out of 100 parts shaded. A decimal square for .27 has 27 out of 100 parts shaded. So .4 > .27

   b. A decimal square for .7 has 7 of the 10 parts shaded. This is also equivalent to a square with 70 out of 100 parts shaded. So .7 = .70

   c. A decimal square for .227 has 227 out of 1000 parts shaded. An equivalent hundredths square has 22.7 out of 100 small squares shaded. This is less than a hundredths square with 35 small squares shaded. So .225 < .35

   d. A decimal square for .1 has 1 of the 10 parts shaded. This is also equivalent to a square with 10 out of 100 parts shaded. A decimal square for .09 has only 9 out of 100 parts shaded. So .09 < .1

2. a. 3/4 = 75/100 = .75    b. 7/100 = .07

   c. 2/3 = .6666...    d. 7/8 = 875/1000 = .875

   e. 4/9 = .44444...   f. 6/25 = 24/100 = .24

3. a. .278 = 278/1000

   b. .353535... = 35/99 because .010101... = 1/99, .020202... = 2/99, etc.

   c. .03 = 3/100

   d. .7326326326... = 1/10 × 7.326326326... = 1/10 × (7 + .326326326...)
      = (1/10 × 7) + (1/10 × .326326326...) = (1/10 × 7) + (1/10 × 326/999)
      = 7/10 + 326/9990 = 6993/9990 + 326/9990 = 7319/9990

4. a. Since there is an 8 in the thousandths place, .878 is closer to .88 than it is to .87. .878 ≈ .88

   b. Since there is a 4 in the hundredths place, .449 is closer to .4 than it is to .5. .449 ≈ .4  (Be careful here when rounding to tenths not to first round to hundredths.)

   c. There is a 6 in the ten thousandths place, so .5096 is closer to .510 than it is to .509.

   d. Each decimal place contains a 6, so no matter what place we round .6666... to its last digit will always round up to 7. To the ten thousandths place it is .6667.

5. a. A decimal square for .7 has 7 of 10 parts shaded. A decimal square for .6 has 6 of 10 parts shaded. Combining them together we have 13/10 shaded. This fills up one whole decimal square and 3/10 of another one. So .7 + .6 = 1.3.

b.  A decimal square for .4 has 4 of 10 parts shaded. Three copies of this would be 12/10 shaded. This fills up one whole decimal square and 2/10 of another one. So 3 × .4 = 1.2.

c.  A decimal square for .62 has 62 of 100 parts shaded. A decimal square for .48 has 48 of 100 parts shaded. The difference between them is 14/100. So .62 − .48 = .14.

d.  A decimal square for .80 has 80 of 100 parts shaded. To divide by .05 we group those 80 small squares into groups of 5 small squares because .05 can be represented by 5 of these hundredths. This results in 16 groups. So .80 ÷ .05 = 16.

6.  a.  Going from left to right, .006 + .38 = .386, then .386 − .2 = .186.

b.  Multiply 62 × 8 to get 496. The product of .62 × .08 should have four digits after the decimal point because we are multiplying hundredths times hundredths to get ten thousandths. .62 × .08 = .0496.

c.  Dividing .14763 ÷ .21 is equivalent to dividing 14.763 ÷ 21. The quotient should be close to 2/3. .14763 ÷ .21 = .703.

d.  Using correct order of operations we perform the multiplication first. .8 × 340 = 272. Then we add, 47 + 272 = 319.

7.  a.  To multiply by 100 move the decimal point two places to the right. 100 × .073 = 7.3.

b.  Since 7 × 6 = 42, 7 × .6 = 4.2.

c.  To divide by 1000 move the decimal point 3 places to the left. 4.9 ÷ 1000 = .0049.

d.  Multiplying by .01 is the same as dividing by 100. Move the decimal point 2 places to the left. .01 × 372 = 3.72.

e.  Think of 15% as 10% plus half of 10%. Since 10% of 260 is 26 and half of 26 is 13, 15% of 260 = 26 + 13 = 39. *[This gives a good mental technique for calculating tips.]*

f.  25% of 36 is the same as 1/4 of 36. Since 9 × 4 = 36, 25% of 36 = 9.

8.  a.  .49 is close to .5, which is 1/2. So .49 × 310 ≈ 1/2 of 310 = 155.
Or, you could also round 310 to 300 and say that .49 × 310 ≈ 1/2 of 300 = 150.

b.  .24 is close to .25, which is 1/4. So .24 × 416 ≈ 1/4 of 416 = 104.
(1/4 of 400 is 100 and 1/4 of 16 is 4).

c.  Since 33% ≈ 1/3, 33% of 60 ≈ 1/3 of 60 = 20.

d.  Since 76% ≈ 3/4, 76% of 40 ≈ 3/4 of 40 = 30.

9.  a.  Since 36% = .36, we multiply 46 × .36 = 16.56 ≈ 16.6.

b.  To find what percent 30 is of 80, we are asking how many hundredths the fraction 30/80 is equivalent to. We can find the answer by converting 30/80 to a decimal by dividing. 30 ÷ 80 = .375, which is 37.5 hundredths. So 30 is 37.5% of 80.

c. Using an estimate first, we see that the answer should be close to 60, because 15 is 1/4 of 60. One way to calculate the exact answer is to use a proportion. If x is the unknown number, then $\frac{15}{x} = \frac{24}{100}$. We can solve this proportion by cross multiplying. This gives 15 × 100 = 24x. So x = 1500 ÷ 24 = 62.5, which means that 15 is 24% of 62.5.

d. The number that is 118% of 125 will be a little bit more than 125 because we are finding more than 100% of the number. We multiply 125 × 1.18 = 147.5.
So 118% of 125 is 147.5

e. 322 will be more than 100% of 230. We can proceed as in part b. Since 322 ÷ 230 = 1.4, 322 is 140% of 230.

10. a. To convert a number from decimal notation to scientific notation, the decimal point needs to be moved to a position where there is one nonzero digit to the left of it. In other words, we want the mantissa to be a number between 1 and 10. Then in order to have the same number we started with we show the number of powers of 10 to multiply the new number by in order to get the original number. This can be found by counting the number of places that the decimal point has moved. 437.8 = $4.378 \times 10^2$.

b. .000106 = $1.06 \times 10^{-4}$.

11. a. $\sqrt{60}$ is irrational because its decimal expression is infinite and non-repeating.

b. The cube root of 27 is rational because it is equal to 3, since $3^3 = 27$.

c. $6\sqrt{8}$ is irrational. $\sqrt{8}$ is irrational, and the product of any rational number times an irrational number is always irrational.

d. $\sqrt{10} + 5$ is irrational because the sum of any rational number and an irrational number is always irrational.

e. $\sqrt{(1/4)}$ is rational because it is equal to 1/2.

f. The cube root of 60 is irrational because its decimal expression is infinite and non-repeating.

12. a. Using a square root key on a calculator, or a guess and check method to get successive approximations we see that $\sqrt{34} \approx 5.8$

b. Here is a possible successive approximation method for approximating the cube root of 18. $2^3 = 8$ and $3^3 = 27$ so it is between 2 and 3. $2.6^3 = 17.576$ and $2.7^3 = 19.683$. Now try $2.65^3$. This is about 18.61, so the cube root of 18 is closer to 2.6 than 2.7. To the nearest tenth it is 2.6.

13. a. Addition is closed on the set of rational numbers. Whenever we add two rational numbers we always get another rational number.

b. Multiplication is not closed on the set of irrational numbers. For example, $\sqrt{12}$ and $\sqrt{3}$ are both irrational numbers, but their product is $\sqrt{36} = 6$, which is rational. (For other counterexamples, take any irrational square root times itself.)

c. Addition is not closed on the set of irrational numbers. For example, $\sqrt{3}$ and $-\sqrt{3}$ are both irrational numbers, but their sum is 0, which is rational.

14. We simplify square roots by removing square factors of the numbers from under the radical.
    a. $\sqrt{405} = \sqrt{(81 \times 5)} = \sqrt{81}\sqrt{5} = 9\sqrt{5}$. (Using a factor tree may help to find factors.)
    b. $\sqrt{24} = \sqrt{(4 \times 6)} = \sqrt{4}\sqrt{6} = 2\sqrt{6}$.

15. a. Call the length of the missing side c. Then by the Pythagorean theorem, $6^2 + 4^2 = c^2$. So $c^2 = 36 + 16 = 52$. Then $c = \sqrt{52}$. ($\sqrt{52}$ can be simplified to $2\sqrt{13}$.)

    b. Call the length of the missing side b. Then by the Pythagorean theorem, $\sqrt{5}^2 + b^2 = 18^2$. So $b^2 = 18^2 - \sqrt{5}^2 = 324 - 5 = 319$. Then $b = \sqrt{319}$.

16. The original price was more than $187, but it was not exactly 15% more than $187. The 15% was calculated as a percent of the original price, not of the sale price. The diagram below shows that $187 is 85% of the original price. So we are asking the question: 187 is 85% of what number? We can find the solution with this proportion $\dfrac{187}{x} = \dfrac{85}{100}$. Dividing $18700 \div 85$ we get $x = 220$. The original price of the coat was $220.

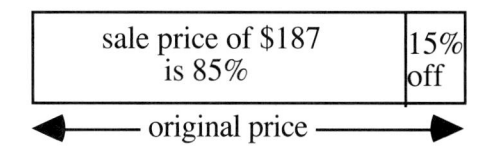

    | sale price of $187 is 85% | 15% off |
    |---|---|

    ◄——— original price ———►

17. We will check the unit price of each glass by finding the cost per ounce. For the 20-ounce glass the cost is $1.50 \div 20 = \$.075$ per ounce. For the smaller glass it is $1.10 \div 16 = \$.06875$ per ounce. The 16-ounce glass is the better buy.

18. The distance from one corner to the opposite corner is the length of the hypotenuse of a right triangle with legs of 60 feet and 30 feet. By the Pythagorean theorem, if we call this length c, then $60^2 + 30^2 = c^2$. So $c^2 = 3600 + 900 = 4500$. Then $c = \sqrt{4500} \approx 67.1$ feet.

19. Since the balance due is over $500 we need to calculate both rates. For the first $500 the amount due is $500 \times .012 = \$6.00$. For the remaining $150, the amount due is $150 \times .008 = \$1.20$. The total amount due is $\$6.00 + \$1.20 = \$7.20$.

20. We are given the ratio of private to public school students, but we are not given either of these amounts. Instead we know the total number of students. But if the ratio of private to public is 3 to 16, then the ratio of public school students to total students is 16 to 19. Call the number of public school students n. Then $\dfrac{n}{18601} = \dfrac{16}{19}$. Solving gives n = 15,664.

# Chapter 7  Statistics

## Section 7.1

1.
   a. There are 7 times as many people in the Army as in the Marines.

   b. The Navy has 400,000. The Marines has 200,000. There are 200,000 more in the Navy.

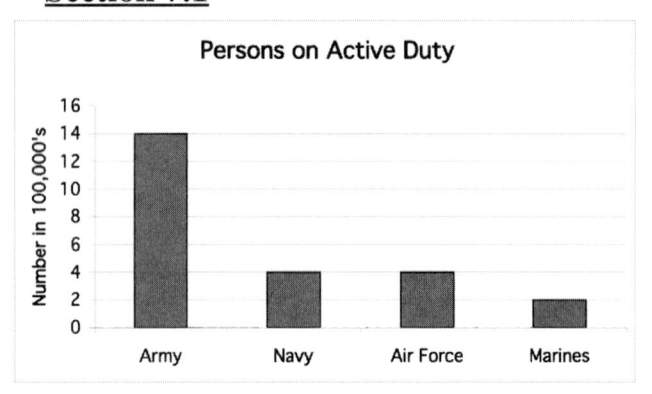

3. a. The tallest bars on the graph are for 1995 and 2000, with a prime rate of 9%.

   b. The prime rate increased 1% in 1994 and 2000, and increased 2% in 1995.

5. a. A protractor is needed for an accurate pie graph. Use central angles from part b.

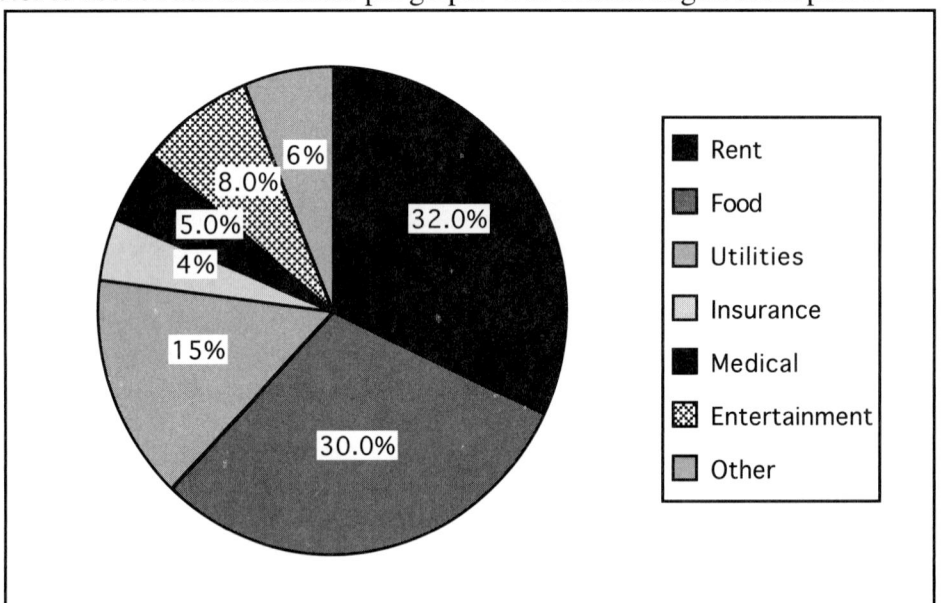

   b. Since a circle is divided into 360°, the central angle for each portion is found by calculating the corresponding percentage of 360°. For rent it is 360 × .32 ≈ 115°, for food 360 × .30 = 108°, for utilities 360 × .15 = 54°, for insurance 360 × .04 = 14.4°, for medical 360 × .05 = 18°, for entertainment 360 × .08 = 28.8°, other 360 × .06 = 21.6°.

## 7.1 Collecting and Graphing Data

7. a. Kansas City has more precipitation than Portland during the summer.

   b. The approximate precipitation for Kansas City is 1+1+2+3+5+4+4+4+5+3+2+2 = 36 inches. The approximate precipitation for Portland is 5+4+4+3+2+2+1+1+2+3+6+6 = 39 inches. Answers may vary due to differences in perception of bar length on the graph.

9. a. The bars for males are longer in bicycle riding, camping, bowling, fishing, basketball, and golf. In these activities males have a greater percentage participation.

   b. In golf the female percentage is only about 2% and the male is about 9%, so the male participation percentage is more than four times as much.

   c. For exercise walking the female percentage is about 20% and the male is about 11%. The female participation is almost twice as much.

11. a. To find which age group has the least difference by race, we look to see which pair of bars is the closest to the same height. The age 20 – 24 category has the least difference.

    b. Since 64.1% of the black women ages 15-19 **do** have early prenatal care, 100 – 64.1 = 35.9% of the black women ages 15-19 **do not** have early prenatal care.

    c. For white women the percentage with prenatal care is increasing for all of the age groups from 15-19 to 30-34. Then it starts decreasing.

13. a. To create this graph we subtract all of the numbers in the given table from 100%.

    b. Since 62.1% of children in poverty under 18 receive public assistance, 37.9% of these children do not receive public assistance. For all children under 18, 26.8% receive public assistance, so 73.2% do not receive public assistance. The difference in percentages for no public assistance is 73.2% – 37.9% = 35.3%.

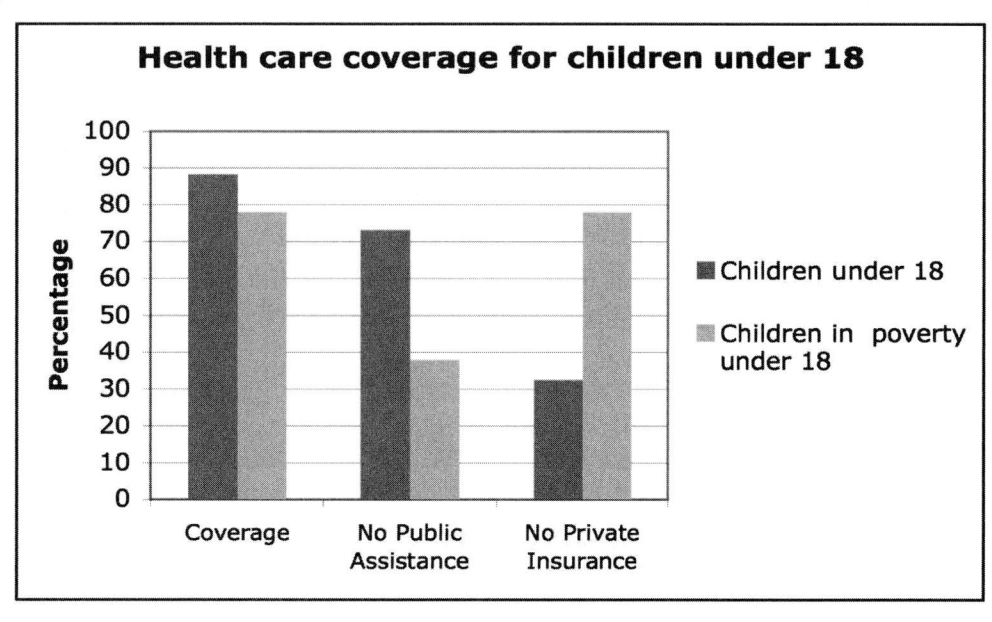

15. a. Hearing impaired is region A, which was 1.3%. Visually handicapped is region D, which was 0.7%. The total spent for these two categories was 2.0%.

   b. 12.4% of the federal funds were spent on programs for the mentally retarded. 1.3% of the same funds were spent on the hearing impaired. 12.4 ÷ 1.3 ≈ 9.5, so there was about 9.5 times as much spent on the mentally retarded as on the hearing impaired.

17. a.

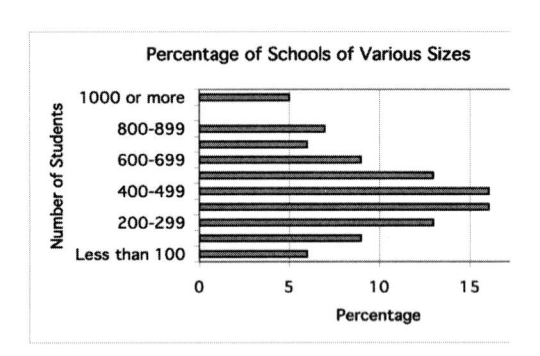

   *Percentage of Schools of Various Sizes* (horizontal bar graph with Number of Students on vertical axis: Less than 100, 200-299, 400-499, 600-699, 800-899, 1000 or more; Percentage on horizontal axis from 0 to 15)

   b. There are three categories that together comprise the schools with from 200 to less than 500 students. Their percents are 13%, 16%, and 16%. A total of 45% of all schools have 200 to 499 students.

19. a. PEOPLE IN AUTO CRASHES BY AGE

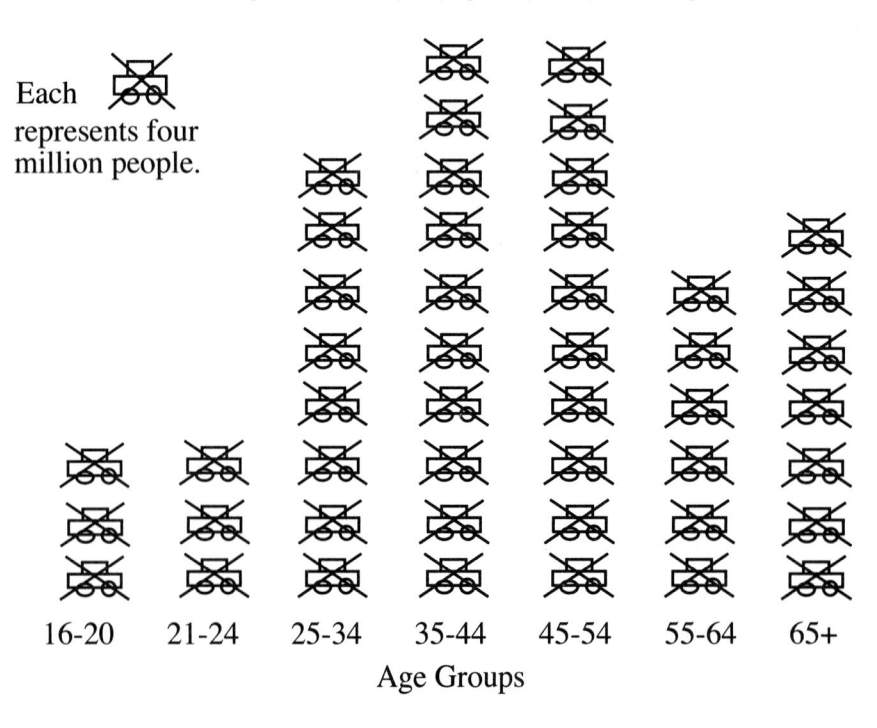

   Each ⊠ represents four million people.

   Age Groups: 16-20, 21-24, 25-34, 35-44, 45-54, 55-64, 65+

   b. There were fewer people in the 16 to 24 age group who had crashes than in the 25 to 34 age group.

## 7.1 Collecting and Graphing Data

21. a.  1999-2000 salaries of classroom teachers by states

```
                      x
                    x x x                                   x
                    x x x       x   x                       x
                    x x x     x x   x                       x
              x   x x x x     x x   x x           x x           x
    x     x x x x x x x x x   x x   x       x x x x x x x x
   ─────────────────────────────────────────────────────────────
    32 33 34 35 36 37 38 39 40 41 42 43 44 45 46 47 48 49 50 51 52 53 54 55 56
```
*Salaries in thousands of dollars*

b. The $38000 range has more states than any other in this line plot.

c. There are 35 out of the salaries in the line plot that are less than $46,000. This is 68.6%.

23. a. Start by forming the stem using the tens digits 4 through 9. Then create the leaves using the units digits.

| | |
|---|---|
| 4 | 9, 6 |
| 5 | 3, 6, 7, 8, 7 |
| 6 | 7, 8, 5, 4, 6, 3, 7, 0, 7, 0, 3, 4 |
| 7 | 3, 8, 9, 1, 4, 7, 0, 1, 2, 8 |
| 8 | 3, 5, 0, 8, 1 |
| 9 | 0, 0 |

b. Seven of the 36 presidents lived 80 years or more. This is about 19.4%.

c. Seven of the 36 presidents did not live 60 years. This is about 19.4%.

25. a. <u>Weights in Kilograms</u> (leaves indicate tenths)

| | |
|---|---|
| 16 | 6 |
| 17 | 0, 2, 3 |
| 18 | 2, 2, 1, 6 |
| 19 | 3, 7, 0, 9, 7, 8, 1 |
| 20 | 2, 5, 3, 1, 6, 4, 0, 8, 1, 6 |
| 21 | 8, 6, 5, 6, 0, 4, 2, 2 |
| 22 | 3, 5, 7, 0, 1, 0 |
| 23 | 8, 4, 1, 2, 0, 0 |
| 24 | 6, 6, 6, 2 |
| 25 | 1, 3, 0 |
| 26 | |
| 27 | 7 |

b. The 20 kilogram stem has the most leaves.

c. The greatest weight is 27.7 kg. and the least weight is 16.6 kg.

27. a. To form a frequency table use one row of the table for the indicated intervals. In the other row indicate the number of data items that occur within each interval.

| Interval | 0-15 | 16-30 | 31-45 | 46-60 | 61-75 | 76-90 | 91-105 | 106-120 |
|---|---|---|---|---|---|---|---|---|
| Frequency | 1 | 14 | 8 | 7 | 3 | 2 | 2 | 1 |

b.

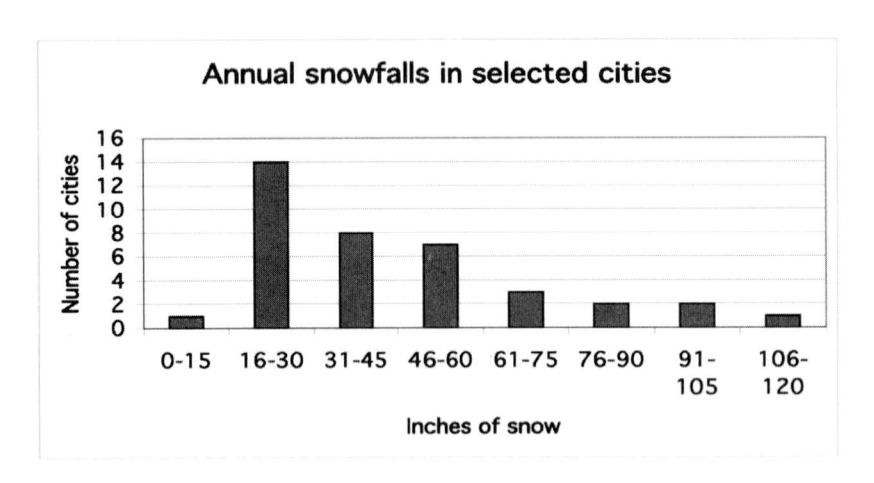

**Annual snowfalls in selected cities**

(Histogram with Number of cities on y-axis, Inches of snow on x-axis; bars: 0-15: 1, 16-30: 14, 31-45: 8, 46-60: 7, 61-75: 3, 76-90: 2, 91-105: 2, 106-120: 1)

c. The 16 to 30 inch interval contains the most cities, with 14.

d. Using the frequency table or the histogram, we see that 3 + 2 + 2 + 1 = 8 cities had snowfalls of more than 60 inches.

29. a. From 1986 to 1988 the line graph is going downhill, so the number of cases of measles was decreasing during those years.

b. From the graph, there were about 7500 cases of measles in 1991. To the nearest thousand this is 5000 more cases than the 2,237 cases in 1992.

c. The graph is highest in 1990. This is the year of the greatest number of reported cases.

d. The greatest two-year decrease occurs between 1990 and 1992. In 1990 there were about 27,500 cases reported. In 1992 the number of cases decreased to 2,237.
So there was a decrease of $27,500 - 2,237 \approx 25,000$ cases.

31. a. The mortgage rate line is lowest in 2003 at a rate of about 6%.

b. The mortgage rate for 1999 was between 7 and 8 percent.

c. The treasury bills line was highest in 1990, at about 7.5%.

d. There are two places where the graphs converge to about 2 percentage points apart. These are in 1998 and in 2000.

33. a.

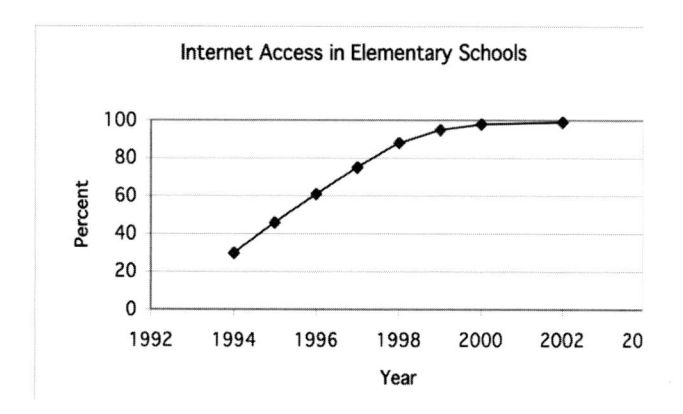

*Internet Access in Elementary Schools* (chart showing percent rising from ~30% in 1994 to near 100% by 2002)

b. From 1994 to 1996 the percent of elementary schools with Internet access approximately doubled.

35. a. One method for drawing a trend line is to locate it so that approximately half of the points are above the line and approximately half are below. For this scatter plot we can draw a line through the point second from left (with midparent height of 163) and the point furthest right (with midparent height of 178). The line through these two points leaves four points above it and four points below it.

b. The median-fit line predicts a daughter height between 158 and 160 cm. for a midparent height of 160 cm. For a midparent height of 174 cm., it predicts a daughter height between 167 and 169 cm.

c. The line predicts a midparent height of 166 to 167 for a daughter height of 163 cm. For a daughter height of 170 cm., it predicts a midparent height between 175 and 177 cm.

37. a. The 1975 black non-Hispanic dropout rate was 27.3% and the 2002 rate was 14.6%. In terms of percentage points this is a decrease of 27.3 − 14.6 = 12.7 percentage points. (Note: It can be confusing (and misleading) when discussing an increase or decrease in statistics that are given in terms of percents. The drop from 27.3% to 14.6% could be called a 12.7% drop, or it could be called a decrease of nearly 50%, since 14.6% is close to half of 27.3%).

b. The percentage point decrease for Hispanic dropouts was 34.9 − 30.1 = 4.8. To the nearest whole number, 12.7 is 3 times as much as 4.8.

c. The scatter plot on the next page shows a positive association between black non-Hispanic and white non-Hispanic dropout rates.

**High School Dropout Rates**

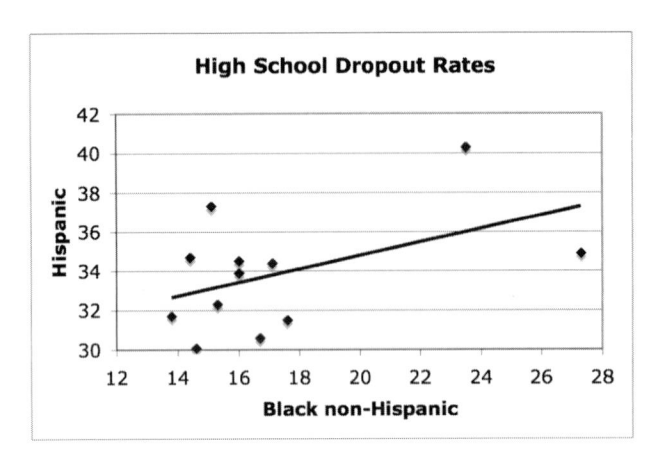

[Scatter plot: Hispanic (y-axis, 30–42) vs Black non-Hispanic (x-axis, 12–28), with upward trend line]

d. A trend line on this scatter plot would predict a Black non-Hispanic dropout rate of between 23% and 24% when the Hispanic dropout rate is 36%.

39. a. The greatest diameter is 8 inches and the oldest age is 42 years.

b. There is a positive association. The trend moves up and to the right.

c. A trend line predicts that a 26-year-old tree will have a diameter of about 5.3 inches and a 32-year-old tree will have a diameter of about 6.5 inches.

d. A tree with a diameter of 9 inches should be about 46 years old, according to the trend line.

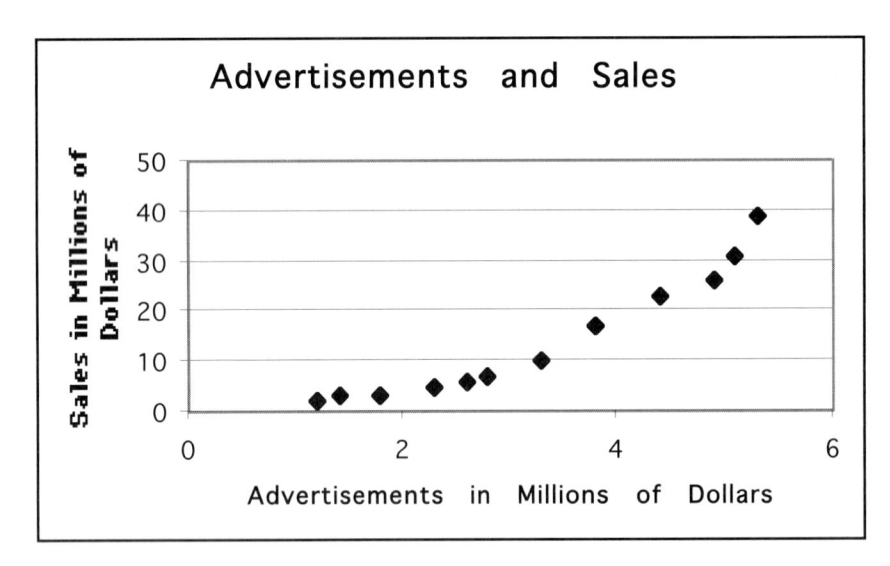

41. a.

**Advertisements and Sales**

[Scatter plot: Sales in Millions of Dollars (y-axis, 0–50) vs Advertisements in Millions of Dollars (x-axis, 0–6)]

b. The shape of this scatter plot fits best with the exponential curve from figure 7.15.

c. An exponential curve fit to these data would show sales of about 12 to 14 million dollars for 3.6 million dollars in advertising.

d. The curve would show about 4 to 4.5 million dollars spent on advertising would result in 20 million dollars in sales.

## Section 7.2

1. a. <u>mean:</u> There are 6 pieces of data. To calculate the mean we can add the data and then divide the sum by 6. 4 + 7 + 6 + 2 + 4 + 5 = 28, and 28÷6 = $4\frac{2}{3}$. The mean is $4\frac{2}{3}$.
   The mean could also be found using a leveling off approach as in Activity 8.1.
   <u>median:</u> To find the median we must first put the data in order from smallest to largest. Then the data is 2,4,4,5,6,7. The median is the middle number. Since there are an even number (6) of numbers we find the mean of the two middle numbers. In this case the median is halfway between 4 and 5 so the median is $4\frac{1}{2}$.
   <u>mode:</u> There are two 4's but no more than one of any other number, so the mode is 4.

   b. <u>mean:</u> There are 8 pieces of data. To calculate the mean we can add the data and then divide the sum by 8. 0 + 1 + 5 + 0 + 2 + 0 + 3 + 1 = 12 and 12÷8 = 1.5. The mean is 1.5.
   <u>median:</u> To find the median we must first put the data in order from smallest to largest. Then the data is 0,0,0,1,1,2,3,5. Both of the numbers in the middle are 1. The median is 1.
   <u>mode:</u> There are more 0's than any other number. The mode is 0.

3. a. A bike shop would be interested in knowing what the most popular bicycle sizes are so that they could have them on hand. The mode would give this kind of information.

   b. For the same reason, the mode is best for describing the typical size of dresses sold in a store. In both this case and in part a. a modification on the mode which gave the top two or three sizes might be even more useful.

   c. The median is usually used to describe the typical cost of homes in a community. If there were a few very expensive homes sold they would skew the mean price up, but would not affect the median.

5. a. For each country the total megawatt capacity and the number of reactors are given. To find the mean capacity for each reactor we need to divide.
   United States: 104,350 ÷ 104 ≈ 1003.4 megawatts
   France:        66,042 ÷ 58 ≈ 1119.4 megawatts
   Japan:         45,907 ÷ 53 ≈ 866.2 megawatts
   Great Britain: 14,240 ÷ 25 = 569.6 megawatts

   b. India has the smallest average megawatt capacity per reactor. 2720 ÷ 14 ≈ 949.

7. a. <u>mean:</u> There are 20 data items. We will compute the sum and divide by 20.
   37+37+49+39+47+40+38+35+46+43+40+47+49+70+65+50+73+49+47+48 = 949.
   949 ÷ 20 = 47.45 The mean number of home runs is 47.45.
   <u>median:</u> When placed in order from smallest to largest, the 10<sup>th</sup> and 11<sup>th</sup> numbers are both 47. The median is 47.
   <u>mode:</u> There are three 47's and three 49's but no more than two of any other number. The modes are 47 and 49.

   b. <u>mean:</u> The sum of the 20 data items is 939. 939 ÷ 20 = 46.95 The mean is 46.95.
   <u>median:</u> When placed in order from smallest to largest, the 10<sup>th</sup> and 11<sup>th</sup> numbers are both 47. The median is 47.
   <u>mode:</u> There are three 40's but no more than two of any other number. The mode is 40.

c. Comparing the numbers in the tables we see that there are 13 years in which the American League's leaders hit at least as many home runs as the National League's.

d. An argument could be made for either league. The medians are the same. The mean is slightly higher for the National League. The American League's top hitter had more home runs than the National League's in ten of the years and they tied in three years.

9. a. The lower quartile is the median of the lower half. It is the left end of the box. In this case it is 65.

b. Approximately one quarter of the scores are in each whisker and in each part of the box. The shaded part of the box runs from 65 to 76. These are about 1/4 of the scores. Since there were 40 scores this is 10 scores.

c. Approximately 3/4 of the scores are above 65. So 30 of the 40 scores were above 65.

d. The interquartile range is the difference between the upper and lower quartiles. We can see it on the box plot by looking at the ends of the box. Here it is 81 – 65 = 16.

11. For the data 52, 61, 67, 75, 79, 81, 82, 83, 90, 93, 96, the median is 81. There are 5 data below the median. The median of these five data, 67, is the lower quartile. The median of the top five data, 90, is the upper quartile. These determine the endpoints of the box and the high and low scores give the endpoints of the whiskers.

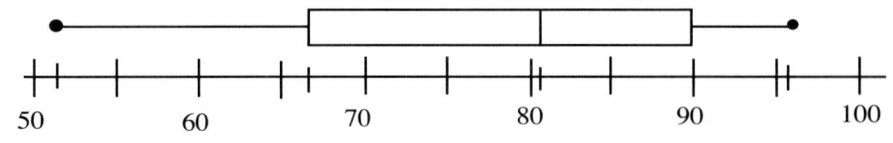

a. The range of these data is 96 – 52 = 44.

b. From the plot we see that the lower half of the data covers a wider range than the upper half. The relatively short whisker on the right shows that the top 1/4 of the data is all between 90 and 96. Other observations can be made as well.

13. a. The 25 data items range from 77.2 to 91.6. They need to be arranged in order first to find the median. It will be the middle, or 13th number. The lower quartile is the mean of the 6th and 7th numbers, 81.25. The upper quartile is the mean of the 19th and 20th numbers, 87.7.

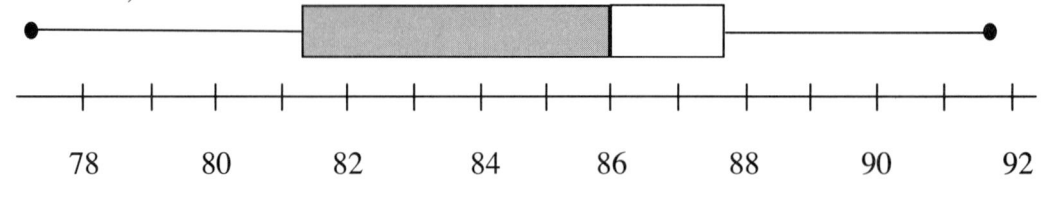

b. The median of 86 indicates that approximately half of these states have less than 86 percent of their students completing high school.

c. The upper quartile of 87.7 indicates that approximately 25% of these states have more than 87.7% of their students completing high school.

d. The lower quartile of 81.25 indicates that approximately 25% of these states have less than 81.25% of their students completing high school.

15. a. The Mirage, Tracer, and Civic appear to have the largest boxes. Using the numbers below to calculate the lengths of the boxes we see that the Tracer's is slightly larger. Its interquartile range is 33 – 25 = 8.

b. The Civic and the Golf both have an upper quartile of 37.

c. The Justy has the lowest median. This is the line inside the box.

d. The Golf has the best ratings. It is comparable to the Civic, but the median is higher for the Golf so it had half of its ratings above 35 while half of the ratings for the Civic were below 33.

17. a. For each network we find the median and the two quartiles. For example, for CBS there are 15 pieces of data, so the median is the $8^{th}$ one, 45.8. The lower quartile is the $4^{th}$ number. It is 44.0. The upper quartile is 48.5, the value of both the $12^{th}$ and $13^{th}$ numbers. In order to compare the three box plots we sketch them all on the same scale.

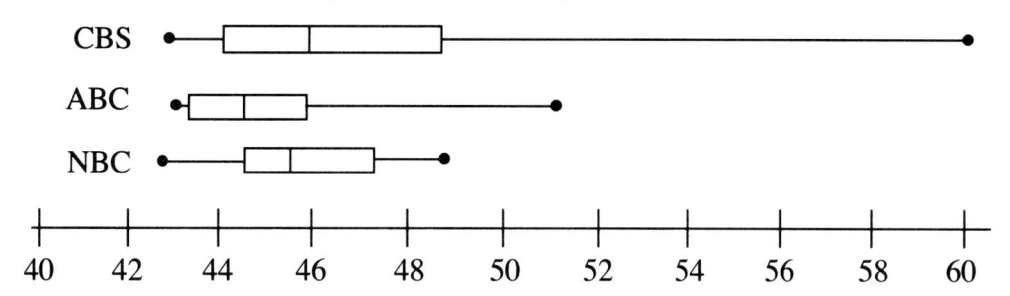

If we look only at the boxes it is close between CBS and NBC. CBS has a slightly higher median and a higher upper quartile. The lower quartile of NBC is better. The higher range for CBS probably gives them the edge. ABC is clearly third.

b. For CBS the interquartile range is 48.5 – 44.0 = 4.5. To locate any outliers we multiply the interquartile range by 1.5. 4.5(1.5) = 6.75. Then add 6.75 + 48.5 = 55.25 and subtract 44.0 – 6.75 = 37.25. No ratings are this low, so there no low outliers. But there is a rating above 55.25. The rating of 60.2 is clearly an outlier. For ABC the interquartile range is 45.5 – 43.4 = 2.55. Multiplying by 1.5 and then adding and subtracting from the quartiles gives a range of about 39.6 to 49.8. Again there is one high outlier, the rating of 51.1. Looking at the plot for NBC we can see that there will be no outliers. Neither whisker is as long as the box. The two outliers indicate that CBS had one program that was very highly rated compared to its other top programs, as did ABC.

19. a. The standard deviation is a measure of how far a typical piece of data deviates from the mean. The data in Set B are more closely concentrated around the mean, so they should have a smaller standard deviation.

b. To calculate the standard deviations find the squares of the differences from the mean for each piece of data, then find the mean of these numbers. The square root of this mean of the squares of the differences from the mean is the standard deviation.

Set A: $(0 - 6)^2 + (2 - 6)^2 + (4 - 6)^2 + (6 - 6)^2 + (8 - 6)^2 + (10 - 6)^2 + (12 - 6)^2 = 112$
The mean of the squares of the differences from the mean is $112 \div 7 = 16$.
The standard deviation is $\sqrt{16} = 4$.

Set B: $(3 - 6)^2 + (4 - 6)^2 + (5 - 6)^2 + (6 - 6)^2 + (7 - 6)^2 + (8 - 6)^2 + (9 - 6)^2 = 28$
The mean of the squares of the differences from the mean is $28 \div 7 = 4$.
The standard deviation is $\sqrt{4} = 2$.

c. The standard deviation for Set B is indeed smaller than for Set A.

21. a. The standard deviation is 9.2, so 2 standard deviations is a distance of 18.4. Since the mean is 72, 2 standard deviations above the mean is 72 + 18.4 = 90.4. There are two measurements above this: 91 and 92. Two out of 55 is approximately 3.6%.

b. The standard deviation is 9.2, so 2 standard deviations is a distance of 18.4. Since the mean is 72, this gives a range of 53.6 to 90.4 that is within 2 standard deviations of the mean. The only data not within this range are 51, 91, and 92. Three out of 55 is approximately 5.5% of the data are either above or below 2 standard deviations.

23. a. The Southern and Western regions each contain 13 states. Remember to arrange the data from lowest to highest before finding the median and quartiles.
For the South the median is 6.9. The lower quartile is the mean of the 3$^{rd}$ and 4$^{th}$ lowest data, which is 6.35. The upper quartile is 7.8.
For the West the median is 8.0, the lower quartile is 6.75 and the upper quartile is 8.2.

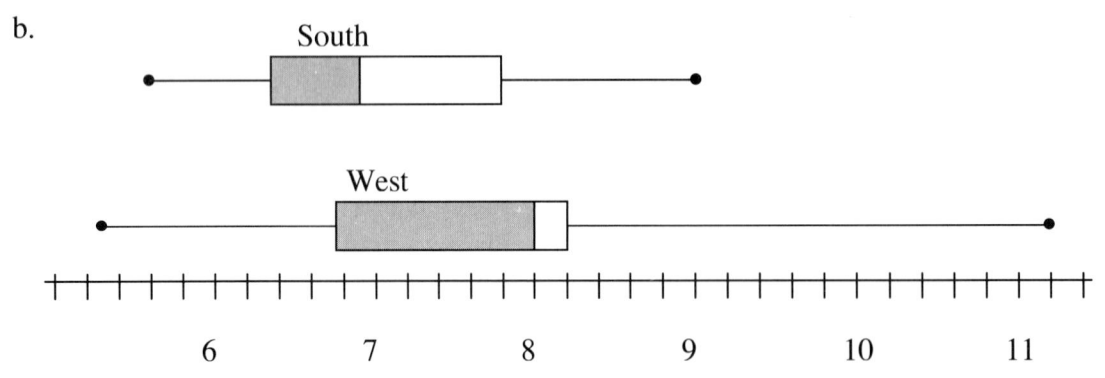

b.

The quartiles and median for the south are all lower than their corresponding measures for the west. The data for the west are more spread out than for the south, especially in the top 25%. (Other observations can also be made.)

c. The median for the west is greater than the upper quartile for the south. So 75 percent of the states in the south spent less per student than the median amount spent per student by the states in the west.

d. The interquartile ranges for the west and the south are the same, so both regions have about the same amount of variability in money spent per student for their midrange states.

25. a. Again arrange the data from lowest to highest before finding the median and quartiles. In the Midwest there are 12 states, so the median is the number between the 6$^{th}$ and 7$^{th}$ pieces of data. In this case they are both 11, so the median is 11. The lower quartile is 10. The upper quartile is 12. For the West the median is 11, the lower quartile is 10.5, and the upper quartile is 15.

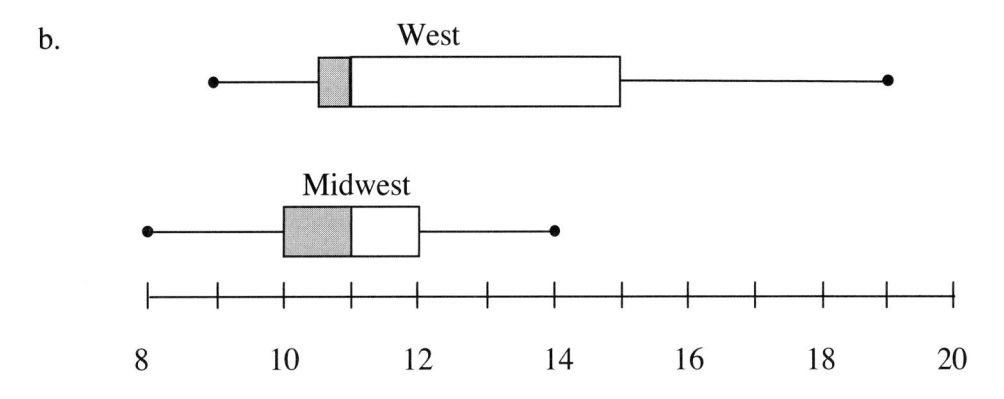

b.

c. The medians for both regions are equal, but the West has higher quartiles. The upper quartile for the West is above the top value for the Midwest, which means that for 25% of the states in the West, the percent of students not finishing high school is above the percentage for all of the states in the Midwest. (Other observations can also be made.)

d. For the Midwest the interquartile range is 12 – 10 = 2. Neither of the whiskers stretch more than 1.5(2) = 3 units beyond the box, so there are no outliers there. For the West the interquartile range is 15 – 10.5 = 4.5. There are no outliers there either.

27. a.	PERSONAL INCOME PER CAPITA (in thousands)

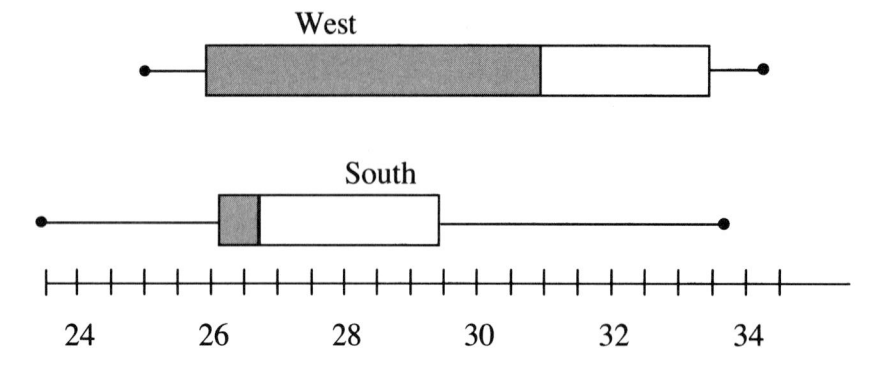

*Box plots for West and South, horizontal axis from 24 to 34.*

b. For the South the interquartile range is 29.4 – 26.1 = 3.3. This represents a difference of $3300 in per capita income between the lower and upper quartiles for states in the South. For the West the interquartile range is 33.45 – 25.9 = 7.55. This represents a difference of $7550 in per capita income between the lower and upper quartiles for states in the West. The smaller interquartile range for the south shows that incomes for the states in the middle 50% of the south are less variable.

c. The box plots show that the upper 50% of the incomes by state in the Western region are above the upper quartile for the South.

d. The median for the South (26.7), is greater than the lower quartile for the West (25.9).

e. The west has more variability in the middle 50% and the south has more variability at both extremes. In at least half of the states in the west the average income is higher than in ¾ of the states in the south.

29. a.	PERSONAL INCOME PER CAPITA (in thousands)

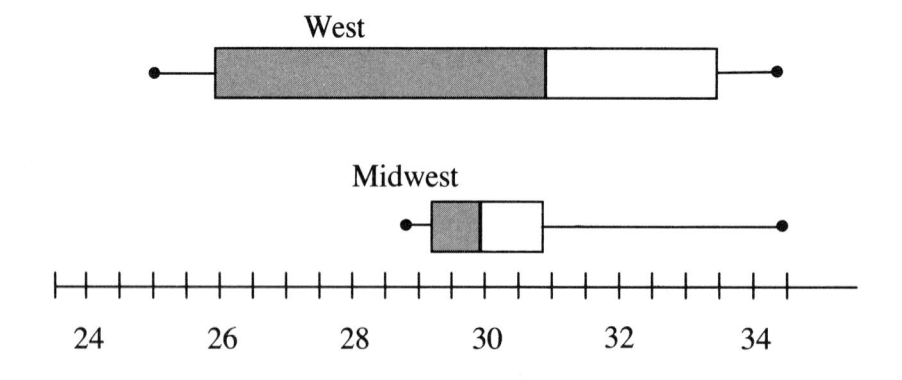

*Box plots for West and Midwest, horizontal axis from 24 to 34.*

b. From the box plot the box for the West looks about 4 to 5 times as long as the box for the Midwest. To verify we find the differences between upper and lower quartiles.
Midwest: 30.85 – 29.2 = 1.65     West: 33.45 – 25.9 = 7.55     7.55 ÷ 1.65 ≈ 4.6.

c. The lower 25% of the states in the West are all below the lowest state in the Midwest.

d. Various conclusions can be drawn. Average incomes in the Midwest are less variable, especially in the lower three quarters. Incomes for all states in the Midwest are above the incomes for the lowest quarter of states in the west. The range of incomes for the top half of the states is almost the same in each region.

31. If 7 scoops of ice cream are to have a mean weight of 45 grams, then their total weight will be 7 × 45 = 315 grams. So far the total weight is 43 + 46 + 44 + 41 + 44 + 45 + 39 = 302 grams. There needs to be an additional 13 grams added to the smallest scoop. This problem can also be solved by observing that 2 – 1 + 1 + 4 + 1 + 0 + 6 = 13 grams need to be added to the smallest scoop to make the average exactly 45 grams.

33. If there are 18 more $5^{th}$ graders than $6^{th}$ graders, and if these classes are in the ratio of 4 to 3, then there are 4 × 18 = 72 in the $5^{th}$ grade and 3 × 18 = 54 in the $6^{th}$ grade. The ratio of $6^{th}$ graders to 7th graders is 2 to 3, so there are 1.5 times as many $7^{th}$ graders as $6^{th}$ graders. There are 54 × 1.5 = 81 in the $7^{th}$ grade.
The mean for all three grades is (72 + 54 + 81) ÷ 3 = 69 students.

35. a. The mean is (4.9 + 6.3 + 5.1 + 6.1 + 5.8 + 6.2 + 5.7 + 6.3 + 6.0 + 5.6) ÷ 10 = 5.8 grams.

b. To calculate the standard deviations find the square of the difference from the mean 5.8 for each piece of data, then find the mean of these 10 numbers. The square root of this mean of the squares of the differences from the mean is the standard deviation.
$(.9)^2 + (.5)^2 + (.7)^2 + (.3)^2 + (0)^2 + (.4)^2 + (.1)^2 + (.5)^2 + (.2)^2 + (.2)^2 = 2.14$
The mean of these squares is 2.14 ÷ 10 = .214. The standard deviation is $\sqrt{.214} \approx .46$ g.

c. The range of weights that are plus or minus one standard deviation is from 5.8 – .46 to 5.8 + .46. This is 5.34 to 6.26. Two are below and two are above this range.
6 out of 10, or 60% of the weights are within one standard deviation of the mean.

37. It is always true that at least 75% of any data is within two standard deviations of the mean. At least 75% of Peter's arrows strike between 14" and 26" away from the center of the target because his mean distance is 20" and two standard deviations either way give a range of 14 to 26. At least 75% of Sally's arrows strike between 8" and 24" away from the center of the target because her mean distance is 16" and two standard deviations either way give a range of 8 to 24.

## Section 7.3

1. a. If the mean is 8975 and the standard deviation is 67, then one standard deviation below the mean is 8975 − 67 = 8908 and one standard deviation above the mean is 8975 + 67 = 9042. 68% of the time the number of heads in 17,950 tosses should be between 8908 and 9042.

   b. Two standard deviations is 67 × 2 = 134. 8975 − 134 = 8841 and 8975 + 134 = 9109. 95% of the time the number of heads in 17,950 tosses should be between 8841 and 9109.

3. a. Number the names for 1 to 9. Randomly select a number from the table and continue on in the table until obtaining two numbers from 1 through 9.

   b. Number the questions from 1 to 60. Randomly select a number from the table and continue on in the table until obtaining 10 numbers from 1 through 60.

5. There are a total of 50 + 80 + 90 + 80 + 100 = 400 students to sample from. A sample of 80 is to be chosen. This is 1/5 or 20% of the students, so in a stratified sample we want to select 20% of the Kindergartners, 20% of the 1st graders, etc. We choose 10 from grade K, 16 from grade 1, 18 from grade 2, 16 from grade 3, and 20 from grade 4.

7. a. A test designed for third graders and given to first graders will have many low scores and a few high scores. The "tail" is to the right, so it is skewed to the right.

   b. A test designed for sixth graders and given to eighth graders will have many high scores and a few low scores. The "tail" is to the left, so it is skewed to the left.

9. a. The day before an exam, most students will spend a large amount of time studying. Some students will not spend much time studying. The graph would be skewed left.

   b. Widths of hand spans of fifth graders should be fairly normally distributed. The graph should be symmetric.

   c. Compared to average sneaker sizes, those of most pro basketball players are very large. The graph would be skewed to the left.

11. a. 68% of the students scored between 400 and 600 because that is the range that is plus or minus one standard deviation from the mean in this normal distribution.

    b. From Figure 8.34 in the text we see that in a normal distribution, 13.5% score between +1 and +2 standard deviations. On this test, that is from 600 to 700. Only 2.35% score between +2 and +3 standard deviations, and about 0.15% score above that. The total scoring above 600 would be 13.5% + 2.35% + 0.15% = 16%.

    c. The percent scoring below 300 is the same as the percent above 700. It is 2.35% + 0.15%, or 2.5%.

13. a. About 2.5% of the scores will be below ⁻2 standard deviations. 2.5% of 50 ≈ 1 student.

    b. About 13.5% of the scores will be from +1 to +2 standard deviations.
       13.5% of 50 ≈ 7 students.

    c. About 13.5% of the scores will be from ⁻2 to ⁻1 standard deviations.
       13.5% of 50 ≈ 7 students.

15. a. From the shape of the graph we can see that it is a normal distribution.

    b.
    | Diameter    | 7 | 8 | 9 | 10 | 11 | 12 | 13 | 14 | 15 | 16 | 17 |
    |-------------|---|---|---|----|----|----|----|----|----|----|----|
    | No. of Trees| 2 | 5 | 8 | 10 | 13 | 26 | 12 | 9  | 8  | 4  | 3  |

    c. From the information above the graph, the mean diameter is 12 and the standard deviation is 2. The trees with diameters between 10 and 14 inches are within one standard deviation of the mean. This is 10 + 13 + 26 + 12 + 9 = 70 out of 100 trees, or 70%.

17. To mark the approximate percentile locations on the graph, start by finding the 50$^{th}$ percentile. This is the median point, where half of the area under the curve is to the left and half is to the right. To find the 25$^{th}$ percentile location, judge where the left half is split in half in terms of area under the curve.

    a. 30% of the measurements are less than or equal to the 30$^{th}$ percentile.

    b. 10% of the measurements are greater than or equal to the 90$^{th}$ percentile.

19. a. Under MATH COMP the national percentile score is 54. This means that 54% of the students nationally scored below this student in mathematics comprehension.

    b. Under MATH COMP the local percentile score is 68. This means that 68% of the students locally scored below this student in mathematics comprehension.

    c. Under READING COMP the national percentile score is 96. This means that this student scored higher in reading comprehension than 96% of the national group.

    d. The mathematics comprehension score is the only one for which the local percentile score is higher than the national percentile score.

    e. If the local percentile score is lower than the national percentile score then it means that compared with the local group the individual did not do as well as he did compared with the national group. In other words, the local group was tougher competition than was the national group.

21. a. In Spelling this student has a national percentile score of 86. This means that 86% of the students nationally scored below this student.

   b. In READING COMP 78% of the students in the local group scored below this student.

   c. This student scored higher than 90% of the national group in Vocabulary.

   d. The top row under MATH APPL reads 32/40. This means that the student answered 32 out of the 40 questions correctly. 32/40 = 4/5 = 80/100. 80% were correct.

23. a. Between the $40^{th}$ and $60^{th}$ percentiles is stanine 5, so the $44^{th}$ percentile is stanine 5.

   b. For Vocabulary the local percentile is 69, a stanine of 6. For Math. Appl. the local percentile is 62, a stanine of 6. For Spelling the local percentile is 83, a stanine of 7. For Language the local percentile is 45, a stanine of 5.

25. The z-score gives the relative position of the score to the mean in terms of standard deviations. To find the z-score, subtract the mean from the score and divide by the standard deviation.
   For Test A:  $2.8 - 3.1 = {^-}0.3$, and ${^-}0.3 \div 2.1 = {^-}0.14$ so the z-score is ${^-}0.14$.
   For Test B:  $5.3 - 6.2 = {^-}0.9$, and ${^-}0.9 \div 1.7 = {^-}0.53$. The z-score is ${^-}0.53$.
   Since ${^-}0.14 > {^-}0.53$ the score for Test A is better.

27. The random numbers contain the digits from 0 through 9. To simulate rolling a die we want to randomly choose from 6 numbers. One way to do this is to let 1 represent rolling a 1, 2 represent rolling a 2, etc. Then we simply ignore the digits 7,8,9, and 0. For example if we start at the beginning of the third row and read straight across then the first ten rolls are 3, 3, 2, 4, 1, 3, 2, 5, 3, 2.

29. a. For the ACT the z-score was $(31 - 17.4) \div 7.8 = 1.74$. For the SAT the z-score was $(582 - 458) \div 117 = 1.06$.

   b. Relative to the other students her performance was stronger on the ACT test because of the higher z-score there. The z-score gives the standing in relation to the mean in terms of standard deviations.

31. The mean is 260° and the standard deviation is 3°, so an acceptable range from 254° to 266° includes exactly plus or minus two standard deviations from the mean (6° either way). Assuming a normal distribution we then have 95% of the crockpots within this range. So 5% of the crockpots can be considered to be defective.

33. The WATS line is overloaded whenever it gets more than 20 calls per minute. This corresponds to the times when the number of calls is more than one standard deviation above the mean. In a normal distribution 16% of the data is more than one standard deviation above the mean, so the line will be overloaded 16% of the time during peak periods.

## 7.3 Sampling, Predictions, and Simulations 153

35. The key figures to use here are the mean loss of 3.3 inches, the standard deviation of 1.4 inches, and the selected measurement of a loss of .3 inches. This measurement has a z-score of $(.3 - 3.3) \div 1.4$, which is approximately $-2.14$. This indicates that this measurement is more than 2 standard deviations away from the mean. That qualifies it as a "rare event" and is evidence that the company's claim is not valid.

37. One way to simulate selecting boxes with one of the five colors of markers would be by rolling a die. Since the die has 6 numbers and we only want to choose from 1, 2, 3, 4, and 5, we could choose to simply ignore the times when a 6 is rolled. Roll the die and record the results, keeping count of the number of rolls until each of 1, 2, 3, 4, and 5 has been obtained at least once. Here are the results of three trials of this simulation.
2 3 2 4 1 3 5   Trial 1 required 7 rolls of 1 die to obtain the target sums 1, 2, 3, 4, and 5.
2 1 2 1 5 5 4 1 3   Trial 2 required 9 rolls to obtain the target sums 1, 2, 3, 4, and 5.
4 4 3 4 4 2 4 1 2 5   Trial 3 required 10 rolls to obtain the target sums 1, 2, 3, 4, and 5.
The mean number of rolls for these 3 trials is 8.6.

   (Note: This is a very small number of trials. Very different results might be obtained in other trails. More data would be more conclusive.)

39. Since there are two equally likely possibilities here, boy or girl, tossing a fair coin is a good simulation. To simulate having children until at least one of each sex is obtained we toss the coin until we have obtained both heads and tails. Here are the results of one such simulation in which ten different "families with at least 1 boy and 1 girl" are created. HT, HT, TTH, TTTTH, HHT, TTH, HT, TH, TH, HT   For these ten trials the mean number of children is $(2 + 2 + 3 + 5 + 3 + 3 + 2 + 2 + 2 + 2) \div 10 = 2.6$.

41. The graph with the vertical scale only going between 191 and 200 gives the impression of an impressive rate of increase in sales. The graph with the vertical scale starting at 0 gives the impression of a much slower rate of growth. From this second graph it looks like the amount of sales has been almost constant over the years. If we use the numbers in the table to calculate the percent increase for some of the years we see that the second graph is a better illustration of reality. For example, from 2004 to 2005 the amount of increase in sales was 2 million, which is a percent increase of about 1%, since 2 is about 1% of 193. The first graph makes it look like it nearly doubled, clearly a false impression.

43. A good strategy for this problem is to count the occurrences of the most frequent letters first. These are: Z 18; A 14; H 12; L 11; O 9. There are 8 each of C, Q, and W. Using the hint, substitute e in place of each Z, t in place of each A, a in place of each H, and o in place of each L. After putting those in, we can see that we should replace O with h, C with i, Q with n, and W with s. The quote begins, "It is remarkable that a . . ."

## Chapter 7 Test

1.  a. Hawaii had the greatest percent of revenue from the state, at 89.2%.

    b. Nevada had the smallest percent of revenue from the state, at 26.8%.

    c. The range of percents of revenue supplied by the state is 26.8% to 89.2%, a range of 62.4 percentage points.

    d. This list is already in order of percent funding from state, so to find the median, we look for the middle of this list. Arizona is the 25$^{th}$ state listed and South Carolina is the 26$^{th}$ and both have 50% funding from the state. So the median for all 50 states is 50%.

2.  a.

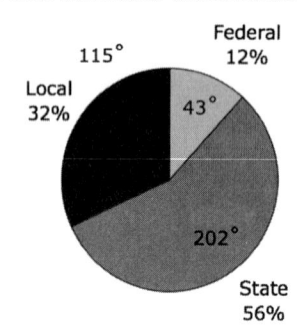

    Revenue Sources for California Schools
    - Federal 12% (43°)
    - State 56% (202°)
    - Local 32% (115°)

    b.

    Revenue Sources for Iowa Schools (bar graph with Federal, State, Local on x-axis and Percents 0–60 on y-axis)

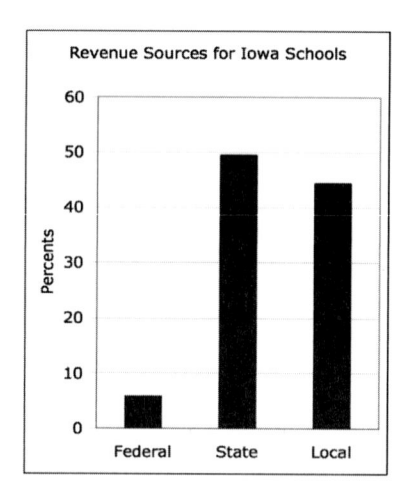

3.  The whole number parts are the stems. Under Federal this ranges from 2 to 15. The leaves represent tenths. To check for accuracy count to see that there are 50 entries in the leaves.

    Percent of Revenue from Federal Govt. for each State

    | Percent | Tenths |
    |---|---|
    | 15 | 0,1 |
    | 14 | 6 |
    | 13 | 2 |
    | 12 | 5,8,9 |
    | 11 | 5,5,8 |
    | 10 | 1,1,5,6,7,8 |
    | 9 | 0,0,1,1,1 |
    | 8 | 1,1,2,2,6,8,8,9 |
    | 7 | 1,7 |
    | 6 | 0,2,4,5,6,7 |
    | 5 | 1,1,5,5,6,6,7,8,9 |
    | 4 | 2,7 |
    | 3 | 9 |
    | 2 | 9 |

4.

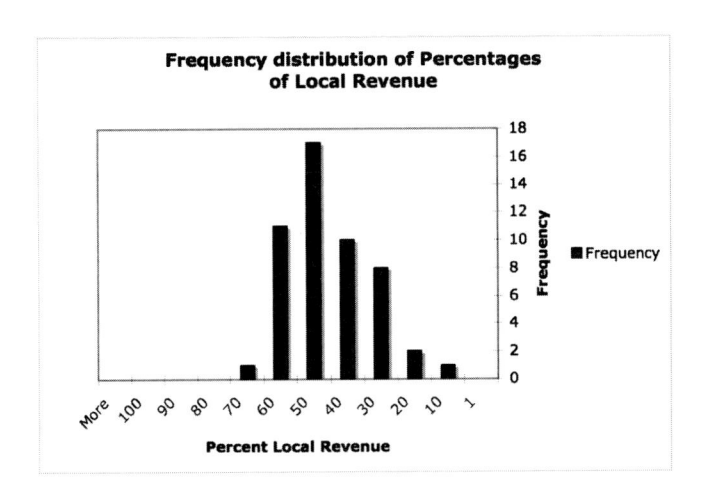

**Frequency distribution of Percentages of Local Revenue**

a. There are 8 states in the interval from 20 to 29.95%.

b. There are 11 of the 50 states in the 50 to 59.95% interval.

c. The 40 to 49.95% interval contains 17 states.

5. a. The percents of local revenue from the 5 states with the highest percent of local revenue are: 57.8%, 58.0%, 58.6%, 58.9%, and 67.5%. The mean of these numbers is found by adding and dividing by 5. To the nearest 0.1%, the mean percent is 60.2%.

b. The percents of local revenue from the 5 states with the lowest percent of local revenue are: 1.7%, 12.6%, 19.8%, 20.3%, and 22.3%. The mean of these numbers is found by adding and dividing by 5. To the nearest 0.1%, the mean percent is 15.3%.

6. In the scatter plot below we draw a trend line. Using this line we see that when the percentage of revenue from local sources is 35, we expect the percentage of revenue from the state to be between 50 and 60 percent.

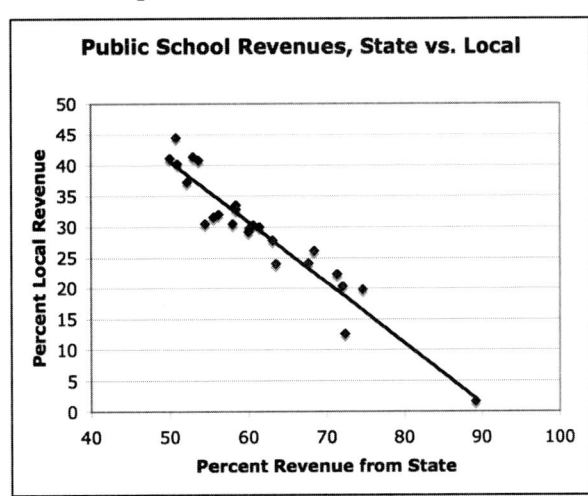

**Public School Revenues, State vs. Local**

7. a. Set B has the greatest mean. We do not need to compute the means to see this. In fact, the mean for Set A is clearly below the minimum value in Set B.

   b. Set A has a greater range. A's range is 20 – 1 = 19 and B's range is 25 – 21 = 4.

   c. Again we do not need to compute to see that Set A has a greater standard deviation. The data are more spread out in Set A.

8. The data ranges from a low of 62 to a high of 97. There are 13 pieces of data, so the median is the 7$^{th}$ one, 73. The lower quartile is 66 and the upper quartile is 89 (halfway between 86 and 92. The interquartile range is 89 – 66 = 23. The range of all of the data is 97 – 62 = 35, so the interquartile range is greater than one half of the range of the data. This can be seen in the box and whisker plot below by noting that the box extends for more than half of the range.

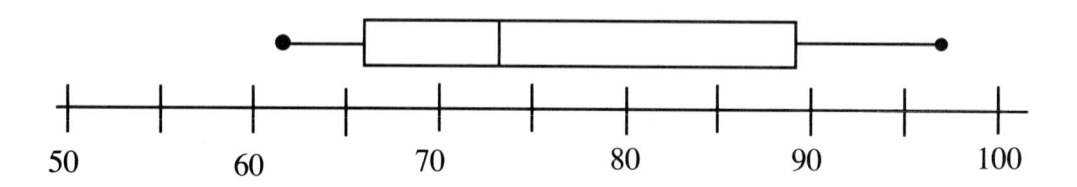

9. a. Mary's z-score for the SAT is calculated by subtracting the mean from her score and dividing the difference by the standard deviation. (520 – 435) ÷ 105 ≈ 0.8. Her z-score for the SAT was 0.8.

   b. For the PSAT her z-score was (56 – 44) ÷ 9.5 ≈ 1.3.

   c. Her mathematics performance was stronger on the PSAT because she did better relative to the rest of those taking the PSAT than she did relative to the rest of those taking the SAT.

10. a. If the pretest is designed as a standard test for mid-year or end of the year 4$^{th}$ graders, then it is likely to be skewed to the right because there will probably be more low scores than high scores. It will have a tail to the right of the few high scores.

    b. For the post-test the distribution will likely be the reverse, skewed to the left.

11. a. In Total Language the student answered 77 out of 93 questions correctly. This is a percent of $77 \div 93 \approx .828 = 82.8\%$.

   b. The national percentile score under Total Reading is 93. This means that 93% scored below this student nationally.

   c. The $60^{th}$ percentile means that this student scored better than 60% of the others locally.

   d. This $4^{th}$ grader reads at a beginning $10^{th}$ grade level, listens at a $9^{th}$ grade level, and has math and language skills at the $5^{th}$ and $6^{th}$ grade levels. The composite is $8^{th}$ grade level.

12. a. In a normal distribution, 68% of the scores are within one standard deviation of the mean. 68% of the third graders scored within plus or minus one standard deviation of the mean.

   b. The standard deviation is 11 and the mean is 74, so a range of 52 to 96 is exactly plus or minus two standard deviations from the mean. In a normal distribution 95% of the scores are within this range.

13. In the first class there are 26 students and the mean score was 68, so there were a total of $26 \times 68 = 1768$ points scored. In the second class there are 22 students and the mean score was 73, so there were a total of $22 \times 73 = 1606$ points scored. The combined classes had 48 students and scored a total of $1768 + 1606 = 3374$ points. The mean for the combined classes is $3374 \div 48 \approx 70.3$ points.

14. Method 1
    The total of the first three tests is $74 \times 3 = 222$. To have a mean score of 78 on 4 tests, a total of $78 \times 4 = 312$ points is needed. The fourth test needs to be $312 - 222 = 90$ points in order to raise the mean to 78.

    Method 2
    If each of the first three tests had 4 more points then the mean would be 78 for these three tests. On the fourth test 78 points plus 12 extra make-up points need to be scored in order to raise the mean to 78. Again, a score of 90 is needed.

# Chapter 8  Probability
## Section 8.1

1. Using the results in figure 8.2 we see that there are a total of 36 possible outcomes of tossing two dice, where we count 3 & 4 as different from 4 & 3.  (Imagine two different colored dice)  There is 1 way to toss a sum of 2, there are two ways to toss a sum of 3, three ways to toss a sum of 4, etc.  The table below shows the probabilities of obtaining each sum.

   | Sum | 2 | 3 | 4 | 5 | 6 | 7 | 8 | 9 | 10 | 11 | 12 |
   |---|---|---|---|---|---|---|---|---|---|---|---|
   | Prob. | 1/36 | 2/36 | 3/36 | 4/36 | 5/36 | 6/36 | 5/36 | 4/36 | 3/36 | 2/36 | 1/36 |

   a. The probability of obtaining a sum greater than or equal to 8 is the sum of the probabilities of obtaining 8, 9, 10, 11, and 12.
   P(sum ≥ 8) = 5/36 + 4/36 + 3/36 + 2/36 + 1/36 = 15/36.

   b. The probability of a sum greater than 4 and less than 8 is equal to
   P(5, 6, or 7) = P(5) + P(6) + P(7) = 4/36 + 5/36 + 6/36 = 15/36.

3. a. P(4, 5, 6, 8, 9, or 10) = P(4) + P(5) + P(6) + P(8) + P(9) + P(10)
   = 3/36 + 4/36 + 5/36 + 5/36 + 4/36 + 3/36 = 24/36 = 2/3.

   b. Since P(8) = 5/36 and P(7) = 6/36, the probability of rolling a 7 is greater.

5. a. There are 7 equally likely numbers to choose from.  Of those seven numbers, four of them, 2, 3, 5, and 7, are prime numbers.  The probability of choosing a prime number is 4/7.

   b. Four of the numbers are less than 5.  The probability of obtaining one of these is 4/7.

   c. Four of the numbers are odd numbers.  The probability of obtaining an odd number is 4/7.

7. a. There are a total of 18 chips in the sample space.  There are 2 red chips.  The probability of selecting a red chip is 2/18 = 1/9.

   b. There are 2 red chips and 5 green chips.  A total of 7 of the chips are either red or green.  The probability of selecting a red or green chip is 7/18.

   c. Of the 18 chips, 16 are not red.  The probability of selecting a chip that is not red is 16/18, or 8/9.  Notice that this is one minus the probability of selecting a red chip.

9. a. For each die, the probability of rolling a 2 is 1 divided by the number of faces on the die.

   b. The probability of rolling a number greater than 3 is 3 less than the number of faces divided by the number of faces.

   |        | Tetrahedron | Cube | Octahedron | Dodecahedron | Icosahedron |
   |---|---|---|---|---|---|
   | P(2)   | 1/4 | 1/6 | 1/8 | 1/12 | 1/20 |
   | P(>3)  | 1/4 | 3/6 | 5/8 | 9/12 | 17/20 |

## 8.1 Single-Stage Experiments

11. a. P(R or G) = 150/360 = 5/12 because there are 360° in the circle and a total of 60° + 90° are in R or G.

    b. P(R or G or B) = 270/360 = 3/4 because there are a total of 60° + 90° + 120° = 270° in R or G or B. Another way to see that the probability is 3/4 is to notice that it is all of the spinner except for Y, which has probability 1/4.

    c. P(G) = 90/360 = 1/4 because there are 360° in the circle and 90° are in G.

    d. P(Y or G) = 180/360 = 1/2 because together they comprise 1/2 of the spinner.

13. a. We will assume that the order of choice does not matter and list all the possible combinations of two of the four letters. They are: AB, AC, AD, BC, BD, and CD.

    b. Three of the six possibilities contain a B chip. The probability that one is a B is 3/6 = 1/2.

    c. Only one of the possibilities has one C and one D. The probability is 1/6.

15. a. Though the order doesn't matter to the customer, we must list BR as distinct from RB in order to get the correct probabilities. The probability of receiving a blue and a red blender is higher than that of receiving two red blenders. The possible outcomes in the sample space are: BB, BR, BY, BG, BW, BP, RB, RR, RY, RG, RW, RP, YB, YR, YY, YG, YW, YP, GB, GR, GY, GG, GW, GP, WB, WR, WY, WG, WW, WP, PB, PR, PY, PG, PW, PP.

    b. There are 36 elements in the sample space. Only one has both blenders blue. The probability of 2 blue blenders is 1/36.

    c. 11 elements have at least one G. The probability of at least one green blender is 11/36.

    d. 25 elements do not contain an R. The probability of not receiving a red blender is 25/36.

17. a. The elements in the set F ∪ H are 2, 3, 5, 6. These are the numbers that are either prime or greater than 4 (or both). P(F ∪ H) = 4/6 = 2/3.
    P(F) = 3/6 = 1/2 and P(H) = 2/6 = 1/3, so P(F) + P(H) = 1/2 + 1/3 = 5/6.

    b. The elements in the set E ∪ F are 2, 3, 4, 5, 6. These are the numbers that are either even or prime (or both). P(E ∪ F) = 5/6.
    P(E) = 3/6 = 1/2 and P(F) = 3/6 = 1/2, so P(E) + P(F) = 1/2 + 1/2 = 1.

    c. Neither of the pairs of probabilities was equal. In order for the probability of the union of two sets to equal the sum of the probabilities of the two sets, the two sets must be disjoint. In other words, they can not have any elements in common.
    *[Can you see why this must be true?]*

19. a. There are 4 aces in the deck of 52 cards, so P(F) = 4/52 = 1/13.

   b. There are 13 hearts in the deck of 52 cards, so P(H) = 13/52 = 1/4.

   c. The set G ∪ H consists of all of the cards that are either spades or hearts. This is half of the deck. P(G ∪ H) = 1/2.

   d. The set E ∪ H consists of all of the cards that are either spades or face cards (or both). There are 13 spades and 3 face cards in each of the other three suits, for a total of 22 cards in this set. P(E ∪ H) = 22/52 = 11/26.

   e. The set G ∩ E consists of spades that are face cards. P(G ∩ E) = 3/52.

21. a. The probability of drawing an ace is 4/52 = 1/13, so the probability of not drawing an ace is 1 − (1/13) = 12/13.

   b. The probability of drawing a face card is 12/52 = 3/13, so the probability of not drawing a face card is 1 − (3/13) = 10/13.

   c. The probability of drawing a spade is 1/4, so the probability of drawing a club, heart, or diamond is 1 − (1/4) = 3/4.

   d. The probability of drawing a black face card is 6/52, so the probability of not drawing a black face card is 1 − (6/52) = 46/52 = 23/26.

23. a. The probability of selecting a club is 1/4. To state the odds of selecting a club we state the ratio of clubs to not clubs. The odds of selecting a club are 1 to 3 because there is one club for every three cards that is not a club.

   b. The probability of selecting an 8 or a 9 is 2/13. The odds are 2 to 11.

   c. The probability of selecting a red card is 1/2. The odds are 1 to 1.
   (There is one red card for every one black card.)

   d. The probability of selecting a spade or a heart is 1/2. The odds are 1 to 1.

25. a. If the probability of living to age 65 is 7/10, then 7 out of 10 people will live to age 65. Another way to say this is that for every 7 people who live to 65, 3 will not. The odds in favor of living to 65 are 7 to 3.

   b. If P(Type O blood) = 3/5, then the odds in favor of type O blood are 3 to 2.

   c. If P(winning the raffle) = 1/500, then the odds in favor of winning are 1 to 499.

   d. If P(rain) = 80% = 8/10, then the odds in favor of rain are 8 to 2, or 4 to 1.

## 8.1 Single-Stage Experiments

27. a. Odds of 10 to 3 for the Yankees winning the pennant can be interpreted to mean that out of 13 tries they would win 10 times and lose 3 times. This is a probability of 10/13.

    b. If the odds of recovery are 1 to 4, the probability of recovery is 1/5.

29. a. If the tack landed point up 19 out of 90 times, then the experimental probability is 19/90. To the nearest hundredth this is $19 \div 90 \approx .21$.

    b. The experimental probability is 194/200 = .97.

31. a. From the table in figure 9.5 for every 10,000,000 births 7,698,698 live to age 60 and 6,800,531 of those live to age 65. So the probability that a 60-year-old person will live to be 65 is $6,800,531/7,698,698 \approx .88$.

    b. From the table 9,781,958 out of 10,000,000 births live to be 12 years old. The experimental probability that a child will live to be 12 is $9,781,958/10,000,000 \approx .98$.

    c. From the table 5,592,012 out of 10,000,000 births live to be 70 years old. The experimental probability that a person will live to be 70 is $5,592,012/10,000,000 \approx .56$.

    d. First we will find the probability that a 28-year-old person will not reach 29. For every 10,000,000 births 9,519,442 live to age 28 and 9,500,118 of those live to age 29. So the probability that a 28-year-old person will live to be 29 is $9,500,118/9,519,442 \approx .99797$. The probability that a 28-year-old person will not reach 29 is $1 - .99797 = .00203$. Out of 7000 28-year-old policy holders the insurance company should expect that about $7000 \times .00203 \approx 14$ will not reach age 29.

33. To use a random number table to simulate tossing a fair coin, we need two equally likely outcomes. One way to do this is to let the odd digits 1,3,5,7,9 represent heads and let the even digits 0,2,4,6,8 represent tails. Reading from the random number table in groups of ten digits we keep track of which groups of 10 contain at least 5 odd numbers. Randomly choose a place to start on the table. For example, we will run 20 trials by starting at the beginning of the 4th row down and reading across through 20 groups of 10 digits. The result for this experiment is that 8 out of the 20 groups of 10 contained at least 5 odds. The experimental probability based on this experiment is 8/20 = 2/5 or 40%.

35. We can use a random number table with odd numbers representing boys and even numbers representing girls. Looking only at blocks of 5 digits in which there are 3 odds and two evens, we count the frequency in which the three odds are consecutive. In one sample of this simulation using the bottom 4 rows of the table in figure 9.6, 3 out of 14 such "families" encountered had 3 boys born in succession. The probability based on this experiment is 3/14. Other ways to simulate this event would be by tossing a coin or drawing slips of paper marked B or G from a hat.

37. If we use a random number table we will want to designate 3 of the digits, say 0, 1, and 2 to represent getting a hit and the other 7 digits to represent not getting a hit. Then we look at groups of five digits and count the frequency of those in which at least 3 of the digits are 0, 1, or 2. In one short experiment using this method starting in the middle of the random number table, two out of twenty groups of five contained at least 3 of these digits. This experiment gave a probability of 2/20 = 10%. Many calculators also have a function to generate random numbers if a table is not available. We could also simulate this situation by placing 10 slips of paper in a sack with three marked H. Each time one is drawn out it needs to be replaced.

39. a. A simulation is not really necessary for this experiment. One could actually toss the coins. Or, if preferred a random number table could be used as described in #33 above, but instead of counting blocks of digits with at least three odds we would count those with exactly three odds. If a computer and software are available, a quicker experiment can be done. Using the Math Investigator dice roll simulation a probability of .32 was obtained.

   b. Since the probability of obtaining exactly 2 heads in a toss of four coins is .375 and the probability of obtaining exactly 3 heads in a toss of 6 fair coins appears to be about .32, it would be reasonable to conjecture that the probability of obtaining exactly 10 heads in a toss of 20 fair coins is less than .5. In fact, the probability seems to be decreasing as the number of coins increases, so a conjecture of less than .3 is also reasonable.

41. a. We want to make the central angles for the blue and purple regions three times larger than the central angles for the green and red regions. The ratios involved here are 3:3:1:1 so we have 8 equal sections to divide the circle into. Since 360 ÷ 8 = 45, the green and red regions should have central angles of 45° (1/8 of the circle) and the blue and purple regions should have central angles of 45 × 3 = 135° (3/8 of the circle).

   b. The ratios involved here are 4:4:1:1 so we have 10 equal sections to divide the circle into. 360 ÷ 10 = 36, so the green and red regions should have central angles of 36° (1/10 of the circle). The blue and purple regions should have central angles of 36 × 4 = 144°.

   c. The ratios involved here are 5:5:1:1 so we have 12 equal sections to divide the circle into. Since 360 ÷ 12 = 30, the green and red regions should have central angles of 30° (1/12 of the circle) and the blue and purple regions should have central angles of 30 × 5 = 150° (5/12 of the circle).

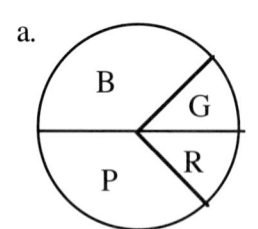

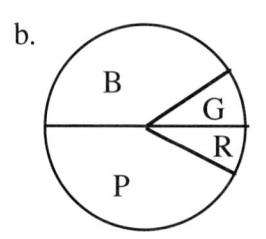

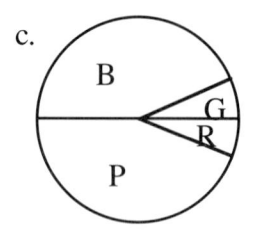

43. We can label the six pairs of roller blades as: G1, G2, G3, D1, D2, D3, representing three good pairs and three defective pairs. The elements of the sample space are all of the possible combinations of two of these pairs. Here is a listing of the sample space.
   G1G2; G1G3; G1D1; G1D2; G1D3;   G2G3; G2D1; G2D2; G2D3;
      G3D1; G3D2; G3D3;   D1D2; D1D3;   D2D3
   There are 15 elements in the sample space and three of them, G1G2, G1G3, G2G3, are favorable. The probability that they will both get a good pair of roller blades is 3/15 = .2

45. One approach is to use a Venn diagram. This diagram uses the given information to show that there are a total of 8 students with either a dog or a cat. So the probability that one of the 24 students has one of these pets is 8/24 = 1/3.

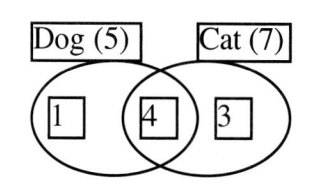

   Another approach is to use the fact that P(D ∪ C) = P(D) + P(C) − P(D ∩ C)
   = 5/24 + 7/24 − 4/24 = 8/24 = 1/3.

47. Let ND be the event of a new division opening and HS be the event of higher sales.
    Then P(ND) = .90, P(HS) = .70, and P(ND ∪ HS) = .75
    By the addition property, P(ND ∪ HS) = P(ND) + P(HS) − P(ND ∩ HS).
    So .75 = .90 + .70 − P(ND ∩ HS).
    Since .75 + 1.60 − .85, we know that P(ND ∩ HS) = .85.
    The probability that she will become vice-president is .85.

49. a. There are a total of 4900 boys and a total of 10,000 students. P(boy) = 4900/10000 = .49
    b. P(girl ∩ District II) = 2300/10000 = .23
    c. P(boy ∪ District III) = (1400 + 2200 + 1300 + 1500)/10000 = 6400/10000 = .64

51. a. P(F ∩ S) = 12%.

    b. P(S ∪ W) = P(S) + P(W) − P(S ∩ W) = 15% + 2% − 0% = 17%.

    c. P(M ∪ S) = P(M) + P(S) − P(M ∩ S) = 25% + 15% − 3% = 37%.
       (Note: M stands for Male, not for married.)

## Section 8.2

1. a. The multiplication rule for probabilities applies here because the two questions are independent events. If the questions are answered by guessing then the probability of getting any one correct is $\frac{1}{2}$. The probability of getting both of the first two questions correct is $\frac{1}{2} \times \frac{1}{2} = \frac{1}{4}$.

   b. The probability of getting the first 5 correct is $\frac{1}{2} \times \frac{1}{2} \times \frac{1}{2} \times \frac{1}{2} \times \frac{1}{2} = \frac{1}{32}$.

   c. The probability of all 10 correct is $(\frac{1}{2})^{10} = \frac{1}{1024}$.

3. a. Again we apply the multiplication rule because one event follows the other. They are independent. P(pink) = 1/2 and P(yellow) = 1/3 so the probability of pink followed by yellow is $\frac{1}{2} \times \frac{1}{3} = \frac{1}{6}$.

   b. P(pink or blue) = 3/4 and P(yellow) = 1/3 so the probability of pink or blue followed by yellow is $\frac{3}{4} \times \frac{1}{3} = \frac{1}{4}$.

   c. P(brown) = 1/4 and P(purple or green) = 2/3 so the probability of brown followed by purple or green is $\frac{1}{4} \times \frac{2}{3} = \frac{1}{6}$.

5. a. The probability of selecting a red marble from the first bowl is 2/3. The probability of selecting a red marble from the second bowl is 1/4. The probability of selecting red marbles from both bowls is $\frac{2}{3} \times \frac{1}{4} = \frac{1}{6}$.

   b. The probability of selecting a red marble from the first bowl is 2/3. If this happens then it does not matter what we select from the second bowl. There is also a 1/3 probability that we do not select a red marble from the first bowl. If this is the case we can still select at least one red marble by selecting red from the second bowl.
   The probability of the event BR is $\frac{1}{3} \times \frac{1}{4} = \frac{1}{12}$.
   So in all there is a probability of $\frac{2}{3} + \frac{1}{12} = \frac{3}{4}$ of selecting at least one red marble.

   c. It is immaterial what happens in the selection from the first bowl if our goal is a yellow marble. There is a 1/2 probability of selecting yellow from the second bowl.
   The probability of selecting a yellow marble is 1/2.

   d.

|   |   | Outcomes | Probabilities |
|---|---|---|---|
| 2/3 → R, 1/4 → G | | RG | 1/6 |
| R, 1/2 → Y | | RY | 1/3 |
| R, 1/4 → R | | RR | 1/6 |
| 1/3 → B, 1/4 → G | | BG | 1/12 |
| B, 1/2 → Y | | BY | 1/6 |
| B, 1/4 → R | | BR | 1/12 |

7.  a. Here are the eight possible birth orders for three children: GGG, GGB, GBG, GBB, BGG, BGB, BBG, BBB. Each of these is equally likely. One of the eight has three girls. The probability that a family of three children has 3 girls is 1/8.

    b. All of the other seven groupings have at least one boy. The probability of at least one boy is 7/8.

    c. Four of the groupings have at least 2 girls. The probability of at least 2 girls is 4/8 = 1/2.

9.  a. Here are the 16 possible outcomes for tossing a coin four times:
    HHHH, HHHT, HHTH, HTHH, HHTT, HTHT, HTTH, HTTT,
    TTTT, TTTH, TTHT, THTT, TTHH, THTH, THHT, THHH.
    Four of these have 3 tails and one head. The probability is 4/16 = 1/4.

    b. There are 11 outcomes above with at least 2 tails. The probability is 11/16.

    c.

    Each path is equally likely

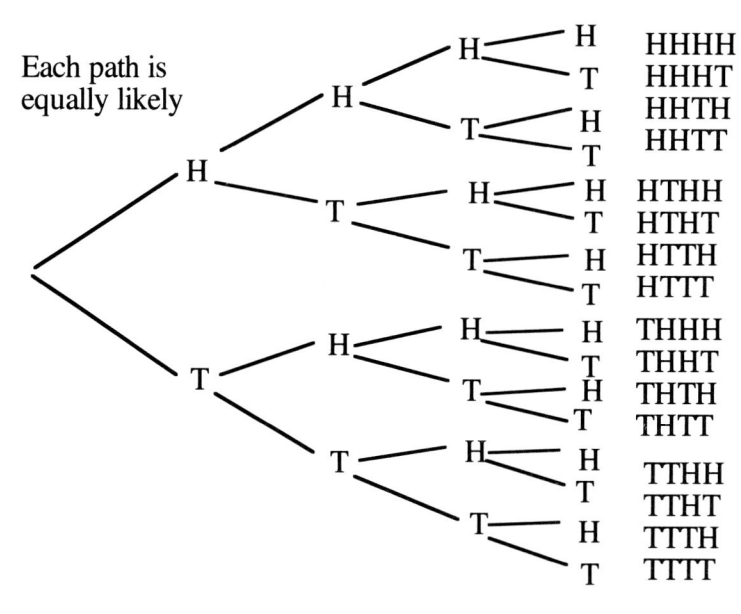

    ```
                    H─── H   HHHH
                 H─┤    T   HHHT
              H─┤   T─── H   HHTH
              │      T   HHTT
              │   H─── H   HTHH
         H─┤  T─┤    T   HTHT
         │    T─── H   HTTH
         │         T   HTTT
    ─────┤
         │    H─── H   THHH
         │ H─┤    T   THHT
         T─┤   T─── H   THTH
              │      T   THTT
              │   H─── H   TTHH
              T─┤    T   TTHT
                 T─── H   TTTH
                      T   TTTT
    ```

11. a. There are a total of 15 marbles in the box and 3 of them are red, so we have a probability of 3/15 or 1/5 of choosing a red marble on the first draw. If we are replacing the marble before drawing again, then the probability is still 1/5 the second time. These are independent events, so the probability of selecting two red marbles is $\frac{1}{5} \times \frac{1}{5} = \frac{1}{25}$.

    b. P(Red) = 1/5 and P(Black) = 7/15, so the probability of selecting a red and then a black is $\frac{1}{5} \times \frac{7}{15} = \frac{7}{75}$.

c. P(Red) = 1/5 and P(Purple) = 5/15 = 1/3, so the probability of selecting a red and then a purple is $\frac{1}{5} \times \frac{1}{3} = \frac{1}{15}$ . *[If we add the three probabilities in parts a, b, and c we get 1/5, which is the probability of selecting a red. Why does this have to be true?]*

13. If we are not replacing the first marble after its selection then on the second selection there is a total of 14 marbles and one less of the color that is chosen first. In part a. the probability of selecting the second red marble is 2/14 = 1/7, so P(Red and Red) = $\frac{1}{5} \times \frac{1}{7} = \frac{1}{35}$ .

    For part b, P(Red and Black) = $\frac{1}{5} \times \frac{1}{2} = \frac{1}{10}$ .

    For part c, P(Red and Purple) = $\frac{1}{5} \times \frac{5}{14} = \frac{1}{14}$ .

15. a. It does not matter what is the past history of the coin flips. If it is truly a fair coin, then the probability of heads on any toss is 1/2, even if the last 100 tosses have all been tails.

    b. Again these events are independent. The probability of rolling a 6 on any turn is 1/6, no matter what the previous rolls have been.

17. a. Two turns of rolling dice are independent events. The first roll has no bearing on the second. The probability of a 7 on either roll is 6/36 = 1/6 (see problem #1 in 9.1). The probability of two 7's in succession is $\frac{1}{6} \times \frac{1}{6} = \frac{1}{36}$ .

    b. If the first ball selected is not replaced then the probability for the second event depends on the results of the first event. These are dependent events. P(Green) on the first selection is 5/8. If the first selection was green, then P(Green) on the second selection is 4/7. The probability of selecting two green balls is $\frac{5}{8} \times \frac{4}{7} = \frac{20}{56} = \frac{5}{14}$ .

19. a. There are 12 equally likely outcomes on the spinner and 11 of them are not fuel. Since they are only making one payment this week the probability of not making a fuel payment is 11/12.

    b. They will not make a fuel payment and not make an electricity payment if they pay any of the other 10 bills. The probability of this event is 10/12 = 5/6.

    c. The two week's payments are independent events, so we use the multiplication rule. P(no fuel payment for two weeks) = $\frac{11}{12} \times \frac{11}{12} = \frac{121}{144}$ .

21. a.  To compute the probability of getting <u>at least</u> one sum of 7 or 11 on 3 rolls of the dice it is easiest to first find the probability of no sum of 7 or 11 and then subtract that result from 1. This technique of using complementary events is valid because the two events
    **1.** at least one 7 or 11 and    **2.** no 7 or 11
    are mutually exclusive and make up the entire sample space. So the sum of their probabilities is 1. P(no 7 or 11 in one toss) = 1 − 6/36 − 2/36 = 28/36 = 7/9.
    P(no 7 or 11 in three tosses) = $\frac{7}{9} \times \frac{7}{9} \times \frac{7}{9} = \frac{343}{729}$ .
    P(at least one 7 or 11 in three tosses) = $1 - \frac{343}{729} = \frac{386}{729} \approx .53$.

   b. First find the probability of never getting a sum of 9 or greater in 5 rolls of a pair of dice.
      P(sum of 8 or less in one roll) = 1 − 4/36 − 3/36 − 2/36 − 1/36 = 26/36 = 13/18 ≈ .72
      P(sum of 8 or less on all 5 rolls) ≈ $(.72)^5 \approx .19$
      P(sum of 9 or greater at least once in 5 rolls) ≈ 1 − .19 = .81

23. a. Since the three wheels are independent we use the multiplication rule.
    P(Bell, Bell, Bar) = $\frac{1}{20} \times \frac{3}{20} \times \frac{1}{20} = \frac{3}{8000} = .000375$ .

   b. P(Plum, Plum, Bar) = $\frac{5}{20} \times \frac{1}{20} \times \frac{1}{20} = \frac{1}{1600} = .000625$ .

25. One way to solve this problem is to use a tree diagram. The diagram below shows that the total probability of taking one of the paths leading to room B is $\frac{1}{9} + \frac{1}{9} + \frac{1}{6} = \frac{7}{18}$ .

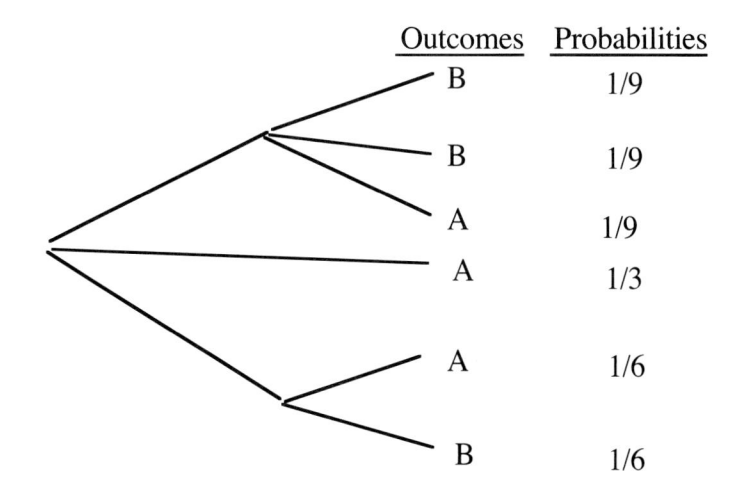

|   | Outcomes | Probabilities |
|---|----------|---------------|
|   | B        | 1/9           |
|   | B        | 1/9           |
|   | A        | 1/9           |
|   | A        | 1/3           |
|   | A        | 1/6           |
|   | B        | 1/6           |

27. a. We are selecting without replacement. The probability that the first bulb is good is 4/5. If a good bulb is chosen the first time then the probability that the second bulb is good is 3/4.
The probability of selecting two good bulbs is $\frac{4}{5} \times \frac{3}{4} = \frac{3}{5}$.

b. Selecting one bad bulb is the complementary event to selecting two good bulbs. The probability is $1 - \frac{3}{5} = \frac{2}{5}$.
Another way to answer these questions is to look at the sample space.
If $G_1$, $G_2$, $G_3$, and $G_4$ are the 4 good flashbulbs and B is the bad bulb, the sample space is:
$G_1G_2$ $G_1G_3$ $G_1G_4$ $G_1B$ $G_2G_3$ $G_2G_4$ $G_2B$ $G_3G_4$ $G_3B$ $G_4B$

29. a. The probability that the second child would be born on April 15 was still 1/365.

b. The probability of four consecutive April 15 births is the product
$\frac{1}{365} \times \frac{1}{365} \times \frac{1}{365} \times \frac{1}{365} = \frac{1}{17748900625} \approx .000000000056$ or $5.6 \times 10^{-11}$.

31. This situation could be simulated by marking one of five slips of paper as a coupon and running trials of drawing four slips out of a sack, with replacement, counting the number of sets of four draws in which the coupon is drawn at least once. Another simulation could use a table of random numbers, designating 2 of the 10 digits as representing the coupon.
We can also find the theoretical probability using the complementary event. The probability of not obtaining at least one coupon in four packages is $\frac{4}{5} \times \frac{4}{5} \times \frac{4}{5} \times \frac{4}{5} = \frac{256}{625}$.
The probability of obtaining at least one coupon is $1 - \frac{256}{625} = \frac{369}{625} \approx .59$.

33. a. To find the expected value of the game we multiply the probability of each possible outcome by its value and then find the sum of all of these products. There are six possible outcomes, each with probability 1/6 and with the values of $1, $2, $3, $4, $5, and $6.
The expected value of the game is
$\frac{1}{6} \times \$1 + \frac{1}{6} \times \$2 + \frac{1}{6} \times \$3 + \frac{1}{6} \times \$4 + \frac{1}{6} \times \$5 + \frac{1}{6} \times \$6$
$= \frac{1}{6} \times (1+2+3+4+5+6) = \$3.50$

b. In order for this to be a fair game a person should pay $3.50 to play. In that case the expected gain or loss is 0.

35. a. The expected value is $\frac{1}{1000} \times \$100 + \frac{1}{500} \times \$50 + \frac{1}{200} \times \$20 + \frac{1}{5} \times \$1 + \frac{1}{100} \times \$10$
$= \$.10 + \$.10 + \$.10 + \$.20 + \$.10 = \$.60$

b. If each ticket costs $1 then this is not a fair game.

37. a. There are 38 equally likely compartments in the roulette wheel, so the probability that the ball will not land on 17 is 37/38.

   b. On page 492 the expected value of a $1 bet on a color was computed to be $ −.05 (a loss of 5 cents). The expected value of a $1 bet on a particular number is $\frac{1}{38}$ ($35) + $\frac{37}{38}$ ($−1)
   ≈ $ −.053 (a loss of about 5.3 cents).
   A more exact calculation of the expected value of the bet on the color also gives a loss of 5.3 cents, so the expected value for each strategy is the same.

39. In the first option the probability of picking a winning ticket is 1/10 = 10%. We can find the probability for the second option by first finding the probability of the complementary event. The probability of not drawing a winning ticket in two draws with replacement from a box of 20 is $\frac{19}{20}$ × $\frac{19}{20}$ = $\frac{361}{400}$ = .9025. The probability of drawing a winning ticket at least once under option 2 is 1 − .9025 = .0975 = 9.75%. You would be slightly more likely to win with option 1.

41. Since the probability that neither engine fails is .98, the probability that either one or the other fails (the union) is 1 − .98 = .02. Let L be the event "failure of the left engine" and R be the event "failure of the right engine". Then P(L) = .02, P(R) = .01, and P(L ∪ R) = .02.
P(L) + P(R) − P(L) × P(R) = .02 + .01 − .0002 ≠ P(L ∪ R)
Hence the events are not independent.

43. Let A be the event of the 1$^{st}$ circuit working and B be the event of the 2$^{nd}$ circuit working. Since the circuits are independent,
P(A ∪ B) = P(A) + P(B) − P(A) × P(B) = .95 + .92 − .95 × .92 = .996.

45. a. There were a total of 250 students and 56 were males who received offers.
    The probability is 56/250 = .224

   b. There were 54 females who did not receive an offer. The probability is 54/250 = .216

   c. 56 + 104 + 54 = 214 were either female or received an offer.
   The probability is 214/250 = .856

47. For each tire there is a 1/12 probability that it will return to the same clock position by chance. The probability that all four tires will return to the same position is
$(\frac{1}{12})^4 = \frac{1}{20736} \approx .00005$

## Chapter 8 Test

1. a. Since there is no replacement we can not repeat letters. Assuming that AB is the same as BA, the sample space has these 15 elements:
   AB, AC, AD, AE, AF, BC, BD, BE, BF, CD, CE, CF, DE, DF, EF

   b. P(A and B) = 1/15

   c. There are 5 elements in the sample space with the A ticket. The probability is 5/15 = 1/3.

2. a. Four of the twelve chips are red. The probability of selecting a red chip is 4/12 = 1/3.

   b. P(red or yellow) = P(red) + P(yellow) = 4/12 + 5/12 = 9/12 = 3/4.

   c. P(not red) = 1 − P(red) = 1 − 1/3 = 2/3.

3. a. The possible outcomes are two green marbles, two orange marbles, or one green and one orange marble. We will symbolize these outcomes as GG, OO, and GO.

   b. From the probability tree below we see that $P(GG) = \frac{2}{3} \times \frac{1}{2} = \frac{1}{3}$.

   c. From the probability tree below we see that $P(OO) = \frac{1}{3} \times \frac{1}{4} = \frac{1}{12}$.

   d. From the probability tree below we see that $P(GO) = \frac{2}{3} \times \frac{1}{2} + \frac{1}{3} \times \frac{3}{4} = \frac{7}{12}$.
   
   The probability for each path is found by multiplying the probabilities for each stage of that path.

   | First Stage | 2nd Stage | Outcomes | Probabilities |
   |---|---|---|---|
   | 2/3 G | 1/2 G | GG | 1/3 |
   |  | 1/2 O | GO | 1/3 |
   | 1/3 O | 3/4 G | GO | 1/4 |
   |  | 1/4 O | OO | 1/12 |

4. a. The odds of the bill's not passing are just the reverse, 5 to 7.

   b. The probability that the bill will pass is 7/12. It is the number of favorable outcomes divided by the total number of outcomes.

5. a. The event E ∪ F takes in all of the names, so P(E ∪ F) = 1.

   b. Evelyn and Eunice are in E ∩ F. P(E ∩ F) = 2/5.

   c. P(E ∪ G) = P(E) + P(G) − P(E ∩ G) = 3/5 + 2/5 − 1/5 = 4/5.

   d. P(F ∩ G) = 1/5, since Frank is the only name in this intersection.

6. a. Since we are selecting the marbles with replacement, the probabilities are the same at both stages. P(red) = 4/6 = 2/3. The probability of selecting a red marble both times is $\frac{2}{3} \times \frac{2}{3} = \frac{4}{9}$.

   b. P(yellow) = 2/6 = 1/3, so P(red then yellow) = $\frac{2}{3} \times \frac{1}{3} = \frac{2}{9}$.

   c. Selecting at least one yellow marble is the complementary event to selecting 2 red marbles. Using the result from part a we get P(at least one yellow) = $1 - \frac{4}{9} = \frac{5}{9}$.

7. a. If we are selecting without replacement then there are only 5 marbles in the sample space for the second selection. The probability of selecting a red marble on the first draw is 2/3. If a red was selected on the first draw then the probability of red on the second draw is 3/5. P(selecting 2 red marbles) = $\frac{2}{3} \times \frac{3}{5} = \frac{2}{5}$.

   b. Without replacement, P(red then yellow) = $\frac{2}{3} \times \frac{2}{5} = \frac{4}{15}$.

   c. Again using complementary events, P(at least one yellow) = $1 - \frac{2}{5} = \frac{3}{5}$.

8. a.

```
                            B ─── B    BBBB
                        B ─┤
                       ╱    G ─── G    BBBG
                      ╱     
                  B ─┤      B ─── B    BBGB
                  │   G ─┤
                  │        G ─── G    BBGG
            B ─┤
           ╱      │        B ─── B    BGBB
          ╱       │   B ─┤
         ╱        G ─┤   G         BGBG
        ╱             │
       ╱              G ─── B    BGGB
      ╱                   ─── G    BGGG
     ┤
      ╲              B ─── B    GBBB
       ╲       B ─┤
        ╲     ╱    G         GBBG
         ╲   ╱
          G ─┤     B         GBGB
              │ G ─┤
              │    G         GBGG
              │
              │    B ─── B    GGBB
              G ─┤       G    GGBG
                   G ─── B    GGGB
                       ─── G    GGGG
```

Each path is equally likely

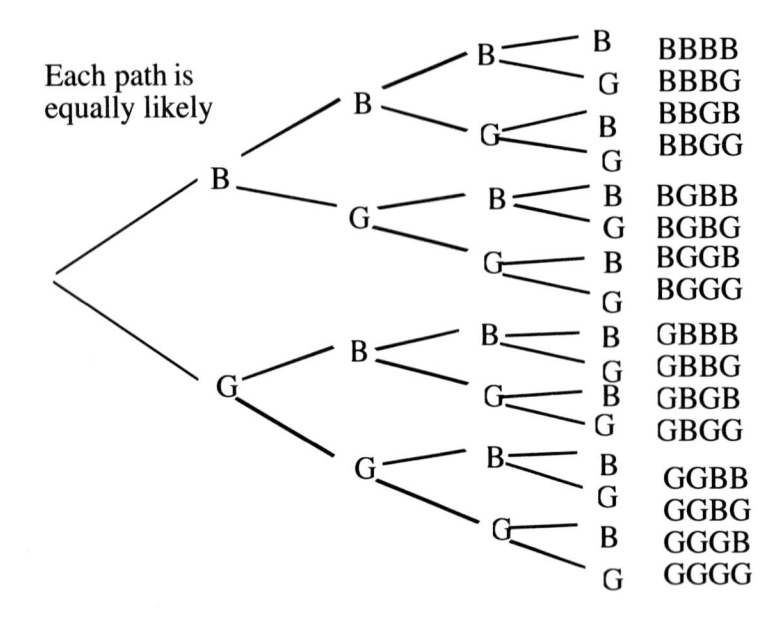

b. There are 16 possible outcomes and each is equally likely, so each has probability 1/16. Six of the 16 outcomes have 2 boys and 2 girls, so the probability of 2 boys and 2 girls is 6/16 = 3/8.

c. Eleven of the 16 outcomes have at least 2 girls, so the probability of at least 2 girls is 11/16.

9. One method of solution is to look at the complementary event. The probability of not choosing a $10,000 envelope on either try is $\frac{5}{7} \times \frac{4}{6} = \frac{10}{21}$, so the probability of choosing at least one $10,000 envelope is $1 - \frac{10}{21} = \frac{11}{21}$.

10. The words "at least" are a clue that using a complementary event might be the easiest approach. The complementary event to the event of obtaining at least one coupon is obtaining no coupons. From any box of cereal the probability of no coupon is 80% or .8. The probability of no coupon from 3 boxes is $(.8)^3 = .512$. So the probability of obtaining at least one coupon from 3 boxes is $1 - .512 = .488 = 48.8\%$.

11. a. To find the expected value for the game we multiply the value for each outcome by its probability. In this case the possible outcomes are 1, 2, 3, 4, 5, or 6 on the die and they each have probability 1/6. The respective values are $1, $2, $3, $2, $5, and $2. The expected value for the game is $\frac{1}{6}(1 + 2 + 3 + 2 + 5 + 2) = \frac{1}{6}(15) = \$3.50$.

b. In order for this to be a fair game it should cost $3.50 to play. Then in the long run the probability would say that the net earnings for either side would be 0.

12. First look at the probability that the system will not fail. The system will not fail if none of the four relays overload. Assuming that the relays are independent in terms of whether or not they overload we can use the multiplication rule for probabilities. The probability of no failure is equal to P(no overload) = $(.99)^4 \approx .96 = 96\%$.
The probability that the system will fail is $1 - 96\% = 4\%$.

13. a. These are independent events, so we apply the multiplication rule. The probability of winning the hurdles and the long jump is $(.9)(.8) = .72 = 72\%$.

    b. The probability of winning all three events is $(.9)(.9)(.8) = .648 = 64.8\%$.

    c. Winning at least one event is the complementary event to winning no events. The probability of winning no events is $(.1)(.1)(.2) = .002 = 0.2\%$. The probability of winning at least one event is $1 - 0.2\% = 99.8\%$.

14. Let F represent the event of access to the first mainframe computer.
    Let S represent the event of access to the second mainframe computer.
    Then P(F) = .9, P(S) = .8, and P(F U S) = .98.
    If the two computers operate independently, then the events are independent and
    P(F U S) = P(F) + P(S) − P(F) × P(S).
    Since $.98 = .9 + .8 - .9 \times .8$, the computers are independent.

15. There are a total of 164 million people represented in the table.

    a. The event described here includes all of the first two rows of the table.
       $15 + 17 + 25 + 31 = 88$, so the probability is $88/164 \approx .54$.

    b. This event includes all of the female column and also the last row of the male column.
       $17 + 31 + 38 + 38 = 124$, so the probability is $124/164 \approx .76$.

# Chapter 9  Geometric Figures
## Section 9.1

1. a. There are many examples of acute angles in the painting. For example the blue triangle at the far right contains two acute angles. Others can be found in the light tan and gray triangles, quadrilaterals, and pentagons just left of and below the center.

   b. Trapezoids can be seen in the inset window frames of two of the buildings.

   c. The upper edges of the two arches at the lower right have right angles.

   d. Several of the figures in the tan and gray region are convex pentagons.

3. The four components of every mathematical system are undefined terms, definitions, axioms, and theorems. <u>Undefined terms</u> are some basic words for which we have an intuitive idea of the meanings, but which we do not define so that we can avoid circularity in our definitions. We then use <u>definitions</u> to give precise meanings to other terms, based on our understandings of undefined terms and previous definitions. <u>Axioms</u> are statements that we assume to be true but do not prove. <u>Theorems</u> are statements which we prove to be true using axioms, definitions, and/or undefined terms.

5. a. To represent a line segment we could use a piece of string pulled tight or a piece of dried spaghetti.

   b. To represent a triangle we could use the end of a triangular prism or a highway yield sign.

   c. A piece of paper or a tabletop could both be models of a plane.
   (None of these physical models can actually be the geometric forms, because of their idealized definitions. Many other models can be chosen to represent the ideas though.)

7. a. None of the angles are acute.

   b. Angles A, C, and D are obtuse angles. Their measures are greater than 90°.

   c. Angles B and E are right angles. Their measures are exactly 90°. This can be verified by holding the corner of a piece of paper up to the angles. Without measuring they may appear to be acute, but this is an optical illusion.

9. a. Because the angles are formed by the intersection of two lines, the sum of any two adjacent angles in this figure is 180°. This means that any pair of adjacent angles here is supplementary. There are four pair of supplementary angles: ∠HOK and ∠KOJ; ∠KOJ and ∠JOI; ∠JOI and ∠IOH; ∠IOH and ∠HOK.

   b. The two pairs of vertical angles are: ∠HOI and ∠KOJ; ∠HOK and ∠IOJ.

   c. The vertical angles are also congruent: ∠HOK and ∠IOJ; ∠HOI and ∠KOJ.

11. a. A chord joins two points on the circle. A diameter is a chord that goes through the center.

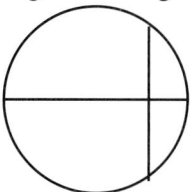

b. A tangent line just touches the circle in one point.

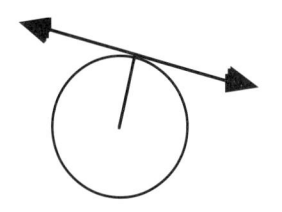

c. If two chords of a circle bisect each other then they are diameters.
(Try drawing two chords bisecting each other that are not diameters.)

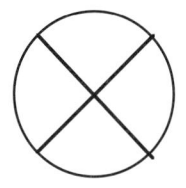

13. Since r and s are parallel lines, angles a, d, f, and h are all congruent, and the other angles b, c, e, and g are their supplements.

   a. The measure of ∠e is 180° − 34.5° = 145.5°.
   b. The measure of ∠h is 34.5°.
   c. The measure of ∠c is 145.5°.
   d. The measure of ∠f is 34.5°.

15. a. At 8 o'clock the hands form a central angle that is one third of the circle. The angle is 360 ÷ 3 = 120°.

   b. From 8 o'clock to 10 o'clock the hour hand moves through one sixth of the circle. The angle it moves through is 360 ÷ 6 = 60°.

   c. When the minute hand has moved through 42° it has moved through $\frac{42}{360} = \frac{7}{60}$ of the circle. Since there are 60 minutes traversed in the circle, 42° represents 7 minutes.

17. a. This curve is simple because it is does not cross itself. It is not closed because the ends do not meet.

   b. This curve is neither simple nor closed.

   c. This curve is closed. It is not simple.

19. a. This region is concave (non-convex). We can find two points in the region for which the segment joining them goes outside the region.

   b. This region is convex. For any two chosen points in the region, the segment joining them lies completely within the region.

   c. This region is concave (non-convex). We can find two points in the region for which the segment joining them goes outside the region.

21. a. Two lines that intersect to form four equal angles must be perpendicular. The angles are 90°.

    b. When three lines intersect to form 6 equal angles the angles must equal 360 ÷ 6 = 60°.

23. If the lines are all parallel there will be no intersections. If they all intersect in a single point then there will be one intersection point. They can not be placed to have either 2 or 3 points of intersection. If four of the lines are parallel and the fifth one passes through each of them there will be four intersection points. The diagrams below show that there are also ways to any number between 5 and 10 intersection points. The possible numbers of intersection points for 5 lines are 0, 1, 4, 5, 6, 7, 8, 9, and 10.

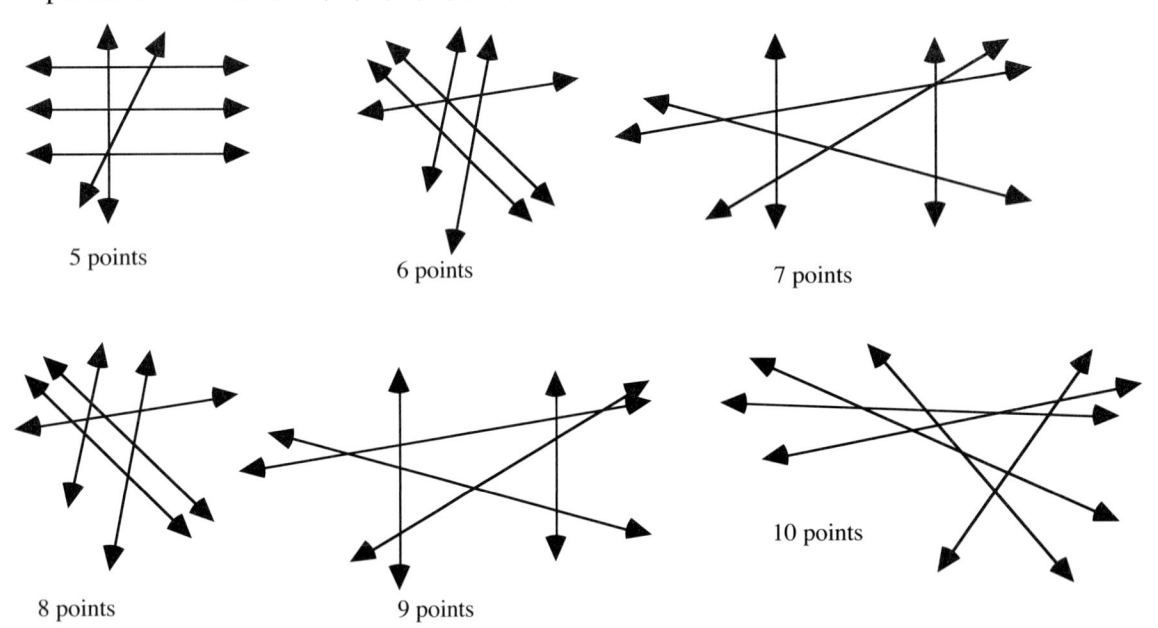

    5 points   6 points   7 points

    8 points   9 points   10 points

25. a. The statement is true. Below is an example of a rectangle, where the diagonals do have the same length, and a nonrectangular parallelogram, in which the diagonals are not equal.

    b. This statement is false. Below is an example of a rectangle where connecting the midpoints does not form another rectangle.

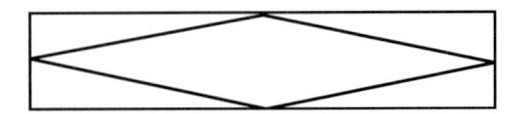

c. The statement is true. Below are two examples.

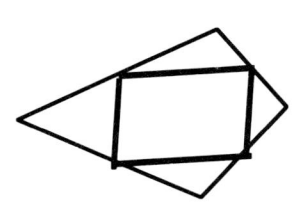

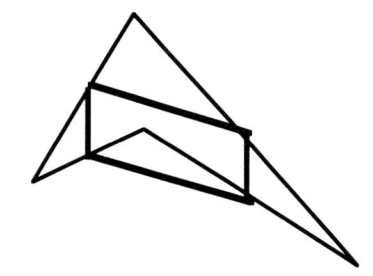

27. If marks are made at one-inch intervals on the pipe, point A can be placed at any of the points marked 1 inch through 13 inches from one end. Since the length from A to B is to be 42 inches, then point B will be at one of the marks from 43 through 55 inches. Putting point A at the 14 inch mark will not work because then point B is at the mark for 56 inches, which is 14 inches from the other end. In this case the two short pieces would both be the same length. Putting point A at one of the marks from 15 through 27 gives the same two short pieces as we get from putting A at 1 through 13. There are 13 different possibilities for obtaining 3 pieces of pipe of different lengths.

29. Number the subscribers 1 through 99. Subscriber number 1 can call each of the other 98 subscribers. Number 2 can call each of the 97 subscribers 3 through 99. (We already counted the connection between 1 and 2. Number 3 can make 96 other connections. Number 4 can make 95 other connections, etc. There are a total of 98 + 97 + 96 + ... + 3 + 2 + 1 possible two-party calls. (This is another application of the staircase problem, or triangular numbers.) The sum of 98 + 97 + 96 + ... + 3 + 2 + 1 is equal to (98 × 99) ÷ 2 = 4851.

31. a. The white path is a simple curve because it does not cross itself. In tracing the path, the pencil mark does not cross over a previous mark.

    b. It is not a closed curve because the two ends do not meet.

33. a. In connecting the points A and B we cross the curve an odd number of times. Hence by the Jordan Curve Theorem the points A and B are not on the same side of the curve.

    b. In connecting the points B and C we cross the curve an odd number of times. Hence by the Jordan Curve Theorem the points B and C are not on the same side of the curve.

    c. In connecting the points D and C we cross the curve an odd number of times. Hence by the Jordan Curve Theorem the points D and C are not on the same side of the curve.

    d. In connecting the points B and D we cross the curve an even number of times. Hence by the Jordan Curve Theorem the points B and D are on the same side of the curve.

## Section 9.2

1. a. The pentagon in the center is surrounded by five hexagons. These hexagons are not regular because some of their interior angles have different measures and their sides are not all the same lengths.

   b. Surrounding the ring of five hexagons is a ring of hexagons and heptagons. Some of these polygons have six sides and some have seven sides.

3. a. A decagon is a 10-sided polygon. If we draw diagonals from one vertex to the seven other non-adjacent vertices we can subdivide a decagon into 8 triangles. The sum of the vertex angles of all of the triangles is the same as the sum of the vertex angles of the decagon. For any triangle the sum of the vertex angles is 180°. So the sum of the vertex angles for a decagon is 8 × 180° = 1440°.

   b. A fifteen-sided polygon can be subdivided into 15 − 2 = 13 triangles. The sum of the vertex angles for a 15-sided polygon is 13 × 180° = 2340°.

5. a. This is not a regular polygon because not all of the interior angles are equal.

   b. This is not a regular (equilateral) triangle because not all of the sides are the same length. The base has length 6. By drawing in the altitudes and using the Pythagorean Theorem we can see that the other two sides have length 5.

   c. This is not a regular polygon because not all of the interior angles are equal. Eight of them are 90° angles and four of them are 270° angles.

7. To find the number of degrees in the central angles we divide 360° by the number of sides.

| Sides | 3 | 4 | 5 | 6 | 7 | 8 | 9 | 10 | 20 | 100 |
|---|---|---|---|---|---|---|---|---|---|---|
| Angle | 120° | 90° | 72° | 60° | 51.4° | 45° | 40° | 36° | 18° | 3.6° |

9. a. If we mark off points on the circle that are the same distance apart as the radius, we will create six equilateral triangles interior to the circle. For any sized circle this procedure will form a regular hexagon.

   b. Connecting every other point we form an equilateral triangle.

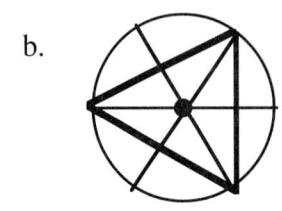

   a. [hexagon inscribed in circle]     b. [equilateral triangle inscribed in circle]

## 9.2 Polygons and Tessellations

11. a. If the central angle is 24° then the total central angle of the regular polygon (360°) has been divided into 360 ÷ 24 = 15 parts. The polygon has 15 sides.

    b. If the central angle is 20° then the total central angle of the regular polygon (360°) has been divided into 360 ÷ 20 = 18 parts. The polygon has 18 sides.

    c. If the central angle is 72° then the total central angle of the regular polygon (360°) has been divided into 360 ÷ 72 = 5 parts. The polygon has 5 sides and is a regular pentagon.

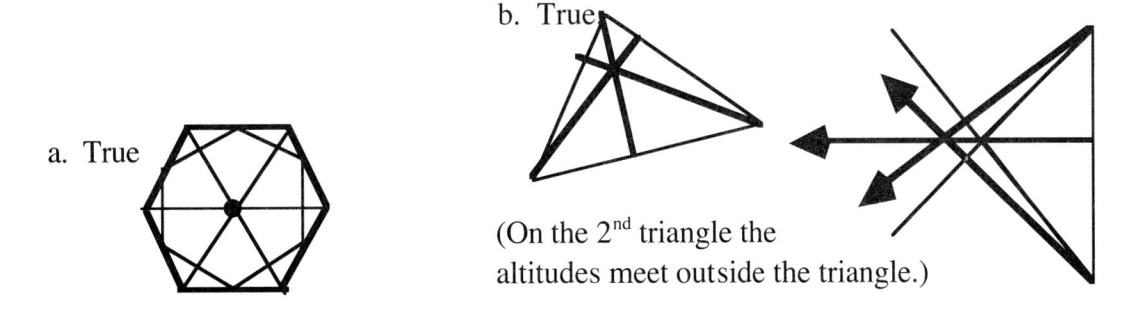

13.
    a. True

    b. True
    
    (On the 2nd triangle the altitudes meet outside the triangle.)

    c. False. Here is a counterexample.

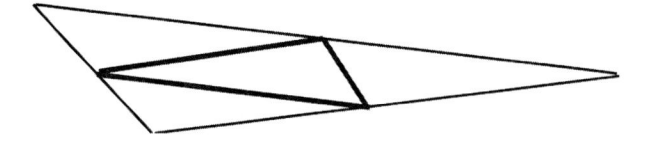

15. The only regular polygons that will tessellate by themselves are the equilateral triangle, the square, and the regular hexagon. So part a. is yes and parts b. and c. are no.
    In order to tessellate the measure of the vertex angle of a polygon must be a divisor of 360.

17. For a figure to tessellate we need to have angles meeting at a point so that the sum is exactly 360°. In any triangle the sum of the three angles is 180°. If we place the triangles so that two copies of each of the three angles meet at a point then we have a total of 360°

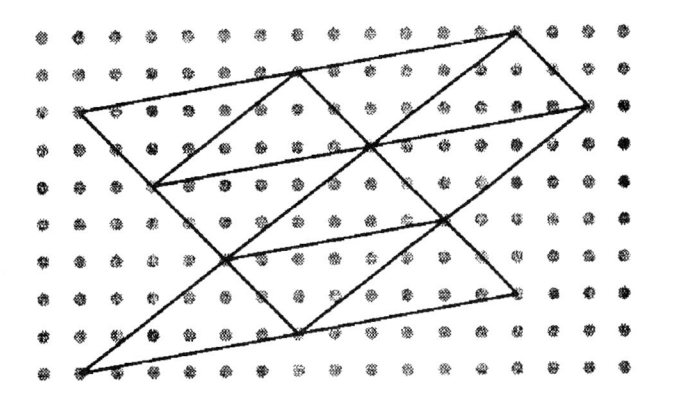

19.

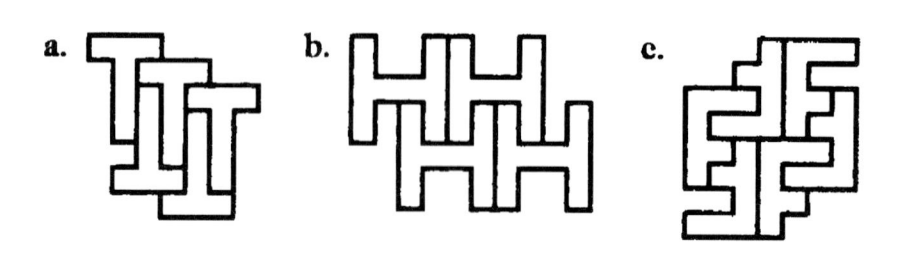

21.

23. a. In a semiregular tessellation each vertex is surrounded by the same arrangement of polygons. This tessellation is not semiregular because we find different arrangements at different vertices. Near the center of this picture are vertices surrounded by a hexagon, two squares and an equilateral triangle. Further out from the center are some vertices surrounded by three equilateral triangles and two squares.

b. This tessellation is semiregular. Each vertex is surrounded by one dodecagon, one square, and one regular hexagon.

25. Here are three methods:
   (1) Draw a line segment one inch long and use a protractor to draw a 150° angle at one end point. Continue drawing sides for the dodecagon and angles of 150°.
   (2) Draw a circle with a compass and draw the central angle for the dodecagon of 30°. Draw a chord between the points where the two sides of the angle meet the circle. Then "pace off" 11 more chords of equal lengths around the circle with a compass.
   (3) Draw a circle and an inscribed regular hexagon, this time using a central angle of 60°. Then draw the perpendicular bisectors of the sides of the hexagon and connect the 12 points on the circle.

27. In order to form a semiregular tessellation we must use at least two of these five polygons. They will not tessellate unless they are chosen in a combination so that angles sum to 360° when they meet at a vertex. Several different combinations will work. Here are two of them.

1.) The vertex angle of the octagon is 135°. Going around each vertex with two octagons and a square gives a total of 135 + 135 + 90 = 360°. Here is a sketch of this tessellation.

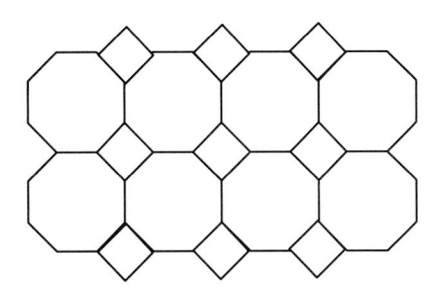

2.) An angle from each of two squares and three equilateral triangles gives 90+90+60+60+60 = 360°. Here is a sketch of this tessellation.

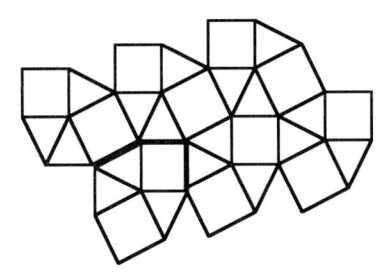

Other semiregular tessellations can also be made from other combinations of these polygons.

29. It takes five toothpicks to make the first pentagon. Each new pentagon shares one side with a pentagon from the previous figure, so each new pentagon after the first one requires 4 more toothpicks. For example, the third figure needed 5 + 2(4) = 5 + 8 = 13 toothpicks to form the three pentagons. The 30$^{th}$ figure requires 5 + 29(4) = 5 + 116 = 121 toothpicks to form the 30 pentagons. There are also other correct ways to describe this pattern and to arrive at the figure of 121 toothpicks for the 30$^{th}$ figure.

31. If we look at the polygons alone without observing the background we may not see anything. Focusing on the white background we see the word LOVE in block letters.

33. First fold a piece of paper in half the long way creating a vertical crease down the center of the paper. Bring the lower right corner of the paper up to meet the crease and fold the left side of the paper down over this so that you create a crease from the lower left corner up to the middle line that is exactly the same length as the bottom of the paper. In the diagram below this is the line segment AC. Unfold the paper and do the same thing on the other side to form the equilateral triangle ABC. We can now find the measurements of 15°, 150°, and 120° in various ways. One way to find each is marked on the diagram below. (The exterior of angle A was bisected to create the 15° angle.)

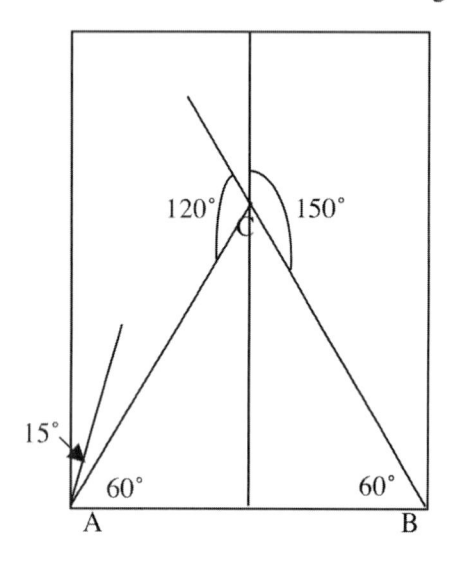

## Section 9.3

1.  a. The top faces have six sides, so they are hexagons.

    b. The bases of these polyhedra are hexagons and the other faces are rectangles. These polyhedra are called hexagonal prisms.

3.  a. This is a polyhedron because each side (face) of this figure is a polygon. Some faces are quadrilaterals and some are triangles.

    b. This is not a polyhedron because it has surfaces that are not polygons.

    c. This is not a polyhedron because it has surfaces that are not polygons.

5.  a. This polyhedron is nonconvex (concave) because we can find points in its interior for which the segment joining them does not completely lie in its interior.

    b. This polyhedron is convex. For every pair of points in the polyhedron, the segment joining the two points lies completely within the polyhedron.

    c. This polyhedron is convex. For every pair of points in the polyhedron, the segment joining the two points lies completely within the polyhedron.

7. a. Looking at the front facing square and going counterclockwise around its upper vertex we see a square, a triangle, a square, and a triangle.

   b. Choose any vertex and note that just three polygons meet at that vertex. Starting with a small square and going counterclockwise we see a square, a hexagon, and a decagon.

9. a. This is a square pyramid. The base is a square and each of the other faces are triangles which share a common vertex at the peak of the pyramid.

   b. This is a cylinder. The two bases are circles. If unrolled, the other face would be a rectangle. Because the upper base lies directly above the lower base it can be called a right cylinder (as opposed to an oblique cylinder).

   c. This is a triangular prism. Two opposite faces (the bases) are triangles. The other three faces are rectangles which join corresponding edges of the bases. Because these are rectangles instead of nonrectangular parallelograms this can be called a right triangular prism (as opposed to an oblique triangular prism).

11. a. This is a cone. It is a right cone because the vertex is directly above the center of the base.

    b. This is an oblique cylinder. The center of the upper base is not directly above the center of the lower base.

    c. This is a right pentagonal pyramid. The base is a pyramid. The peak of the pyramid is directly above the center of the base.

13. a. Face GHIJKL is the upper base. The lower base ABCDEF is parallel to it.

    b. Face GLFA is parallel to face IJDC.

    c. Faces ABHG and ABCDEF meet along the edge AB. The dihedral angle formed by these faces is the same measure as angle GAF (or angle HBC). These are right angles, so the measure of the dihedral angle is 90°.

    d. The measure of the dihedral angle between ABHG and BCIH is the same as the measure of angle ABC (or angle GHI). Since ABCDEF is a regular hexagon this angle is 120°.

15. a. For an arctic region, close to the pole, a plane projection would be best suited. A conic projection could also be suitable if the vertex of the cone were placed above the equator rather than the pole.

    b. For a large region near the equator such as the Western hemisphere between 20°N and 20°S a cylindrical projection would show the least distortion.

    c. For the United States a conic projection would show the least distortion.

17. a. One obtuse angle is formed by the hands of the clock. One can also see obtuse angles in some roof lines at the left of the picture near the chimney.

b. Many rectangles can be found. Some are windowpanes.

c. There are semicircles forming the upper arches of several of the windows.

d. Some of the windowpanes are squares. The ones below the clock and next to the clock appear to be square.

e. The faces of the roof of the tower form isosceles triangles.

19. a. Just east of Madagascar is 20°S, 60°E. This is the point antipodal to 20°N, 120°W because it is just as far south of the equator as the first one is north of the equator and the longitudes, 60 and 120, are supplementary and in opposite hemispheres. The longitudes need to be 180° apart because 360° brings you all the way around the globe.

b. The antipodal point for (30°S, 80°E) is the point (30°N, 100°W). This point is in the United States.

21. a. Assuming that the picture shows a plane intersecting a regular right pentagonal prism and that the plane is parallel to the bases, then the cross section is a regular pentagon.

b. A plane intersecting a right cylinder, with the intersection perpendicular to the bases, will produce a cross section that is a rectangle.

23. Each of the other four pyramids is formed by using one of the four faces of pyramid FHCA as its base and one vertex of its cube as the opposite peak of the pyramid. For example, using face ACH and vertex D we can form pyramid ACHD. The other pyramids are ACFB, CFHG, and AFHE.

25. To make these sketches it may be helpful to construct the figures with some cubes and view them from each of the three perspectives.

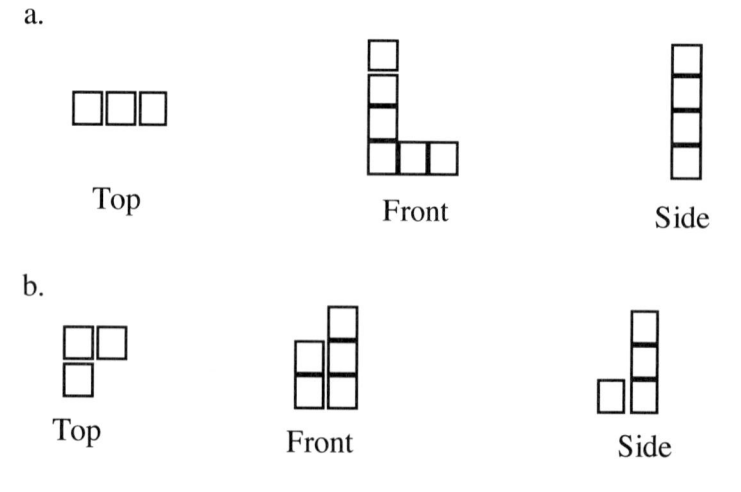

    a.

27. a. The polyhedra numbered 1 through 7 are pyramids. They each have a single base, with lateral faces rising up from each edge of the base and meeting at a common vertex.

   b. A dodecahedron has 12 faces, which are all regular pentagons. Polyhedron number 26 appears to be a dodecahedron.

29. Euler's formula, $F + V - 2 = E$, says that for any polyhedron the number of edges is two less than the sum of the number of faces and the number of vertices.

   a. Using Euler's formula, $7 + 7 - 2 = E$, so there are 12 edges. Polyhedron number 6, the hexagonal pyramid, has 7 faces, 7 vertices, and 12 edges.

   b. Using Euler's formula, $16 + V - 2 = 24$, so $V = 10$ and there are 10 vertices. Polyhedron number 12 has 16 faces, 10 vertices, and 24 edges.

   c. Using Euler's formula, $F + 5 - 2 = 8$, so $F = 5$ and there are 5 faces. Polyhedron number 4, the square pyramid, has 5 faces, 5 vertices, and 8 edges.

31. On September 15 the storm was at its eastern-most point. The coordinates were about (32°N, 48°W). Its coordinates on September 23 were about (31°N, 65°W). On September 30 its coordinates were about (35°N, 76°W).

33. If the cross section is to be the same for both figures, then imagine finding a place to slice the completed structure of the two figures (one inside the other) so that viewed head-on the slice appears to be of just one object. In other words there needs to be a slice of the completed structure where the inner figure is packed tight within the outer one, with no space in between. Here are two possibilities for doing this.
   (1) A cylinder that just fits inside a square prism. Both objects have vertical cross sections that are the same.
   (2) A sphere that just fits inside a cylinder. Both objects have the same horizontal cross section.

186       *Chapter 9 Geometric Figures*

35. a. Roll up a rectangular sheet of paper and tape the opposite edges.

    b. Cut out and roll up a sector of a disc and tape the radii. See the figure below at left.

    c. Hold the right cylinder from part a at an angle, dip the ends at the same angle into a liquid, and cut off the moistened part. Cutting along the taped edges produces the pattern shown above in the figure on the right.

37. a. The first two pictured pentominoes can be folded into an open-top box. In each case the center square becomes the bottom of the box.

    b. Here are four more pentominoes that can be folded into an open-top box. There are also two others not shown here that can be folded into an open-top box. In the diagram below the square marked B is the bottom of the box.

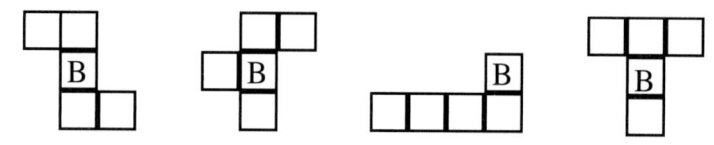

39. a. The table in #36 shows the number of vertices, faces, and edges for each of the five regular polyhedra. The cube has 6 faces and 8 vertices while the octahedron has 8 faces and 6 vertices. This allows us to reverse the roles of vertices and faces and move from cube to octahedron and vice versa. The dual relationship is possible because of this connection between numbers of vertices and faces.

    b. The dodecahedron and icosahedron are also duals. The dodecahedron has 12 faces and 20 vertices while the icosahedron has 20 faces and 12 vertices.

    c. The tetrahedron is its own dual. It has 4 faces and 4 vertices. Placing a vertex at the center of each face of a tetrahedron and joining them forms another tetrahedron.

41. a. The three cylinders are all actually the same size. We perceive the one in the back of the picture as larger than the one in the front because the lines drawn in the picture add perspective which makes us view this as a three-dimensional object. It really is a two-dimensional object and there is no "front" or "back" cylinder.

    b. Using the corner of a piece of paper as a guide, we see that angles 2 and 3 are right angles, angle 1 is obtuse, and angle 4 is acute. So angle 1 is the largest angle. If we perceive this two-dimensional drawing as a three-dimensional cube, then we want to see angles 1 and 4 as the right angles.

## Section 9.4

1. a. Here are five pairs of symmetric objects. The left hedge and right hedge; the column at the top of the fortress; the center arch and itself; the arches on the left and the arches on the right; the surface of the pool and itself.

   b. The structure at the top left of the fortress and the rectangular window on the right side of the fortress don't have images. There are other windows and openings on the fortress that do not have images for the vertical plane of symmetry. The people in the picture are also examples.

   c. The rectangular windows on the fortress and the surface of the pool have horizontal lines of symmetry.

3. a. There are 2 lines of reflection, a horizontal line through the center and a vertical line through the center. There are 2 rotation symmetries. We can rotate either 180° or 360° and see the same picture.

   b. There are 5 lines of reflection, one line for each point of the star. There are also five rotation symmetries. We can rotate 72°, 144°, 216°, 288°, or 360°.

5. a. This polygon has 4 lines of symmetry and 4 rotation symmetries.

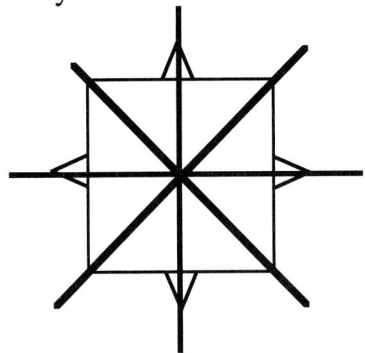

   b. This polygon has no lines of symmetry and has 2 rotation symmetries.

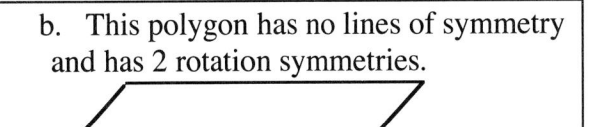

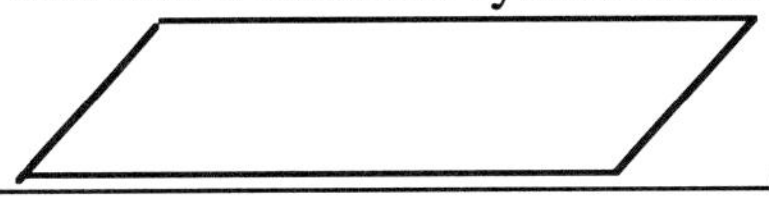

   c. The regular pentagon has 5 lines of symmetry and 5 rotation symmetries.

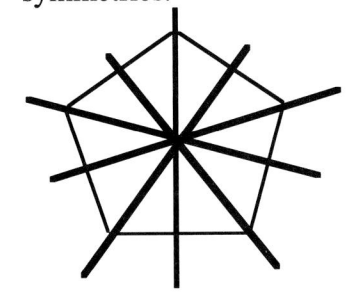

   d. This polygon has 1 line of symmetry and 1 rotation symmetry.

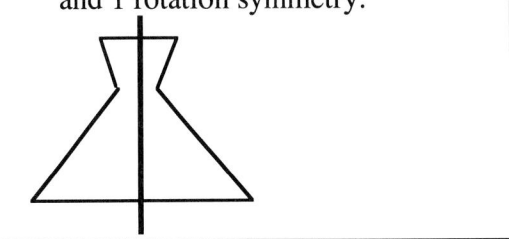

   According to Birkoff's formula, polygon **c** has the highest rating, 10, and polygons **b** and **d** have the lowest rating, each with only 2 symmetries.

7. If we view the dashed vertical line as a mirror and reflect the point A, its image will not be on the figure. It is to the right of the figure. In order for the dashed line to be a line of symmetry it would need to reflect every point back onto the figure. So it is not a line of symmetry.

9. Figures a and c do not have any lines of symmetry, so using the Mira on these images will never show coinciding images. In figure b, placing the Mira along the line through the lower left vertex at a 45° angle to the base will give coinciding images and show that this is a line of symmetry.

11. a. H, X, O, and I have 2 lines of symmetry. All of these have both vertical and horizontal lines of symmetry. The X and O also have other lines of symmetry.

    b. N, S, and Z have 2 rotation symmetries but no lines of symmetry.

13. Figure **d** has no line of symmetry.

    a. This figure has 2 lines of symmetry and 2 rotation symmetries, 180° and 360°.

    b. This figure has 1 line of symmetry, a vertical line.

    c. This figure has 3 lines of symmetry and 3 rotation symmetries, 120°, 240°, and 360°.

    d. This figure has no lines of symmetry, but has 2 rotation symmetries, 180° and 360°.

15. There are many possible correct answers. One example for each part is shown below.

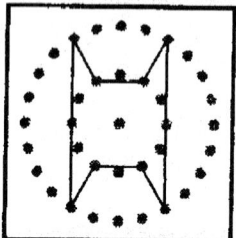

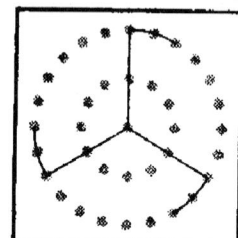

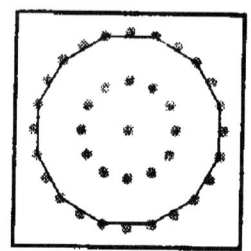

17. To complete the symmetric figures it is helpful to use the grid lines to find the placement of the vertices and the edges of each polygon. The completed figures are shown below.

    a.

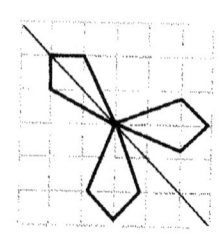

    b.

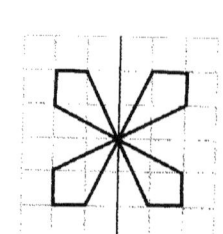

19. a. Using a vertical line through the center of the lampshade as an axis of symmetry, there are 16 rotation symmetries for the lampshade with 16 panels. There are also 16 planes of symmetry, eight of them going through centers of panes of glass and 8 of them going through edges of the glass panes.

   b. This object has only one plane of symmetry, a vertical plane cutting through its center from front to back.

21. a. To visualize an axis of symmetry imagine a stiff wire running through the object to represent an axis. Rotate the object around this axis. If the object appears exactly the same before it has been rotated a complete 360°, then this is an axis of symmetry. Each of the given figures has a vertical axis of symmetry through their centers, but only the right cylinder, equilateral prism, sphere, and cube also have at least one <u>horizontal</u> axis of symmetry.

   b. For the right cylinder there are two rotation symmetries, 180° and 360°. For the equilateral prism there are two rotation symmetries, 180° and 360°. For the sphere there are an infinite number of rotation symmetries. If the axis is a horizontal line through the center, any rotation keeps the appearance of the sphere the same. For the cube there are four rotation symmetries, 90°, 180°, 270°, and 360°.

   c. To see planes of vertical symmetry, imagine a mirror slicing through the object. If it is a plane of symmetry, then the reflection of one side will coincide exactly with the other side. The rectangular pyramid has exactly two planes of vertical symmetry, each of them going through the midpoints of opposite sides of the base. The equilateral prism has exactly three planes of vertical symmetry. The cube has exactly four planes of vertical symmetry.

23. If we consider this object to be a perfect sphere on a pentagonal base, then there are five planes of vertical symmetry, one vertical plane going through each of the vertices of the base, and five rotation symmetries about a vertical axis through the center of the sphere. If we consider the carvings on the sphere to be part of the object then there are probably no symmetries. (It is difficult to determine without seeing the whole object.)

25. a. There are four rotation symmetries; 90°, 180°, 270°, and 360°. There are also four lines of symmetry, two going through midpoints of opposite sides and two going through opposite vertices of the large square.

   b. There are eight rotation symmetries; 45°, 90°, 135°, 180°, 225°, 270°, 315°, and 360°. There are no lines of symmetry.

27. a. There are six rotation symmetries. They can be seen by counting the protruding arms.

   b. There are five rotation symmetries.

29. a. The JPMorganChase Bank logo has 4 rotation symmetries but no lines of symmetry.

   b. The Texaco logo has only one vertical line of symmetry.

   c. The Volkswagen logo has only one vertical line of symmetry.

   d. The Chrysler logo has five rotation symmetries but no lines of symmetry.

   e. The Shell Oil logo has only one vertical line of symmetry.

31. a. The Candelabrum has one vertical line of symmetry.

   b. The Trinity symbol has three rotation symmetries and three lines of symmetry.

   c. The Jerusalem cross has four rotation symmetries and four lines of symmetry. There are vertical and horizontal lines of symmetry and also two diagonal lines of symmetry.

   d. The Byzantine cross has only one vertical line of symmetry.

   e. The Star of David has six rotation symmetries, but no lines of symmetry. There are no lines of symmetry because of the weaving under and over in this symbol.

33. If the plumbing paths are symmetric in the two apartments then the distance between the two terminal points is the length of the horizontal line joining them. In the apartment shown the terminal point is a distance of 6 + 10 + 5 = 21 feet from the dividing wall. If we disregard the thickness of the dividing wall then the two terminal points are 21 + 21 = 42 feet apart.

35. One difficult part of this problem is in understanding the rules of the game. Read them carefully before proceeding. It may help to play an imaginary round and total your score. We score 5 bonus points for landing in the following pairs of symmetrical cups:
    2 and 4; 4 and 6; 6 and 8; 8 and 2; 1 and 5; 3 and 7.
    There is one ambiguous point, which is clarified by the parenthetical note at the end. Landing in cups 1 and 5 scores 6 points for the numbers and 5 bonus points for symmetry about the horizontal line through 3 and 7. Does this pair also qualify for symmetry about the vertical line? According to the note the answer is yes, so a total of 6 + 5 + 5 = 16 points are awarded for landing in cups 1 and 5. Landing in cups 3 and 7 gives 7 + 3 + 5 + 5 = 20 points. If the five bean bags land in cups 1, 3, 5, 7, and 8, then the total score is 16 + 20 + 8 = 44 points. Can we do any better than this? Yes. The best score is from landing in cups 3, 4, 6, 7, and 8. There are bonus points of 10 for 3 and 7, 5 for 6 and 8, and 5 for 4 and 6. The total score is 3 + 4 + 6 + 7 + 8 + 10 + 5 + 5 = 48 points.

37. a. For a vertical line of symmetry we need to have the three dots in the left column exactly match the three dots in the right column. There are various ways this can be done.

   b. For a horizontal line of symmetry the top row and the bottom row must be the same. It does not matter what is in the middle row, because a horizontal line will split these dots exactly in half anyway.

   c. There are only four ways to arrange the dots to have 180° rotation symmetry, but no lines of symmetry.

   ● ○      ● ○      ○ ●      ○ ●
   ○ ○      ● ●      ○ ○      ● ●
   ○ ●      ○ ●      ● ○      ● ○

# Chapter 9 Test

1. a. Since it has 6 sides the hexagon is shown in figure (i).

   b. The parallelogram is shown in (iii). Both pairs of opposite sides are parallel.

   c. The trapezoid is in (ii). It has one pair of parallel sides.

   d. The equilateral triangle is in (i). All three sides are the same length.

   e. Since it has 5 sides the pentagon is shown in figure (vi).

   f. The isosceles triangle is in (v). It has two sides the same length.

2. a. A pentagon has 5 sides. Nonconvex means that we can find two points interior to the pentagon whose connecting segment goes outside the pentagon.

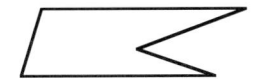

   b. A simple curve does not intersect itself and a closed curve starts and stops at the same point. Here are two examples of simple closed curves.

c. A decagon has 10 sides. Here is a convex decagon.

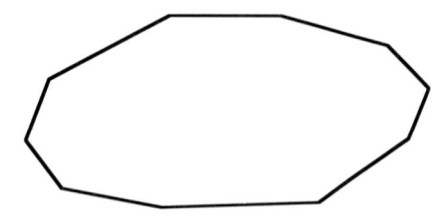

d. For a closed curve that is not simple we start and stop at the same point, and include at least one crossing point. Here are two examples. To draw the second figure without lifting the pencil, starting at point A draw the circle and then draw the ellipse, starting and ending also at point A.

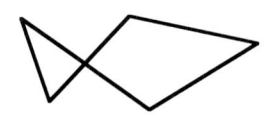

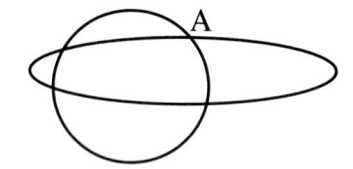

3. Looking at the angles in the interior of the polygon, ∠C and ∠E are acute because they are less than 90°, angle D is a reflex angle because it is greater than 180°, angle B is a right angle because it is exactly 90°, and angle A is obtuse because it is greater than 90° and less than 180°.

4. a. True. Every square is a quadrilateral with two pairs of parallel sides and 4 right angles. (But not every rectangle is a square.)

   b. True. We can draw a right triangle having no sides the same length.

   c. False. Some parallelograms do not contain right angles.

   d. True. Every rectangle is a quadrilateral with two pairs of parallel sides.

   e. False. All equilateral triangles have all three angles of 60° measure.

5. a. To find the measure of the central angle of a regular polygon we divide 360° by the number of sides in the polygon. In the case of the regular octagon we get 360 ÷ 8 = 45°.

   b. To find the measure of a vertex angle of a regular polygon we look at the number of triangles the polygon can be subdivided into. For the hexagon there are four. So the sum of the vertex angles is 180(4) = 720. Each vertex angle has measure 720 ÷ 6 = 120°.

   c. The exterior angle is supplementary to the vertex angle. For a regular pentagon each vertex angle has measure 180(3)/5 = 108°. Each exterior angle then has measure 180 − 108 = 72°.

6. a. This is not a regular polygon because not all vertex angles are equal.

   b. This is not a regular polygon because not all sides are equal.

   c. This is not a regular polygon because not all sides are equal.

   d. This is not a regular polygon because not all vertex angles are equal. Some are acute and some are reflex angles.

7. In order to tessellate the measure of the vertex angle of a polygon must be a divisor of 360. The only regular polygons that will tessellate by themselves are the equilateral triangle with a vertex angle of 60°, the square with a vertex angle of 90°, and the regular hexagon with a vertex angle of 120°. So part c. is yes and parts a. and e. are no. Any isosceles triangle will tessellate and any quadrilateral will tessellate, so the answer for both parts b and d is yes.

8. A semiregular tessellation is a tessellation of two or more noncongruent regular polygons in which each vertex is surrounded by the same arrangement of polygons. The vertex angles that we have to work with are the square (90°), the triangle (60°), and the octagon (135°). We need a combination of these to add to 360°. We could use two octagons and a square, but there is no way to use all three of these polygons to form a semiregular tessellation.

9. a. This is a right pentagonal pyramid. Its base is a pentagon and the vertex joining all of the other edges is directly above the center of the base.

   b. This is a right rectangular prism (otherwise known as a box).

   c. This is a right hexagonal prism. It is right because the two hexagonal bases are directly above and below each other.

   d. This is an oblique cylinder. Its two circular bases are offset.

   e. This is a right cone.

   f. This is an oblique triangular pyramid.

10. A polyhedron is a figure in 3-dimensional space whose faces are polygons. A polyhedron can not have a curved surface. Prisms, pyramids, cubes, and dodecahedra are polyhedra, but spheres and cones are not because they have curved surfaces.

11. We will use Euler's formula, which relates the number of vertices, faces, and edges of any polyhedron in this way: $F + V - 2 = E$.
    a. An icosahedron has 20 triangular faces and has 30 edges. So solving $20 + V - 2 = 30$ for V we get $18 + V = 30$ and $V = 12$. There are 12 vertices in an icosahedron.

    b. For a polyhedron with 14 faces and 36 edges, $14 + V - 2 = 36$, so $12 + V = 36$, and $V = 24$. There are 24 vertices.

12. a. An equilateral triangle is one example of a figure with three lines of symmetry.

    b. Here are two examples of figures with two rotation symmetries but no lines of symmetry.

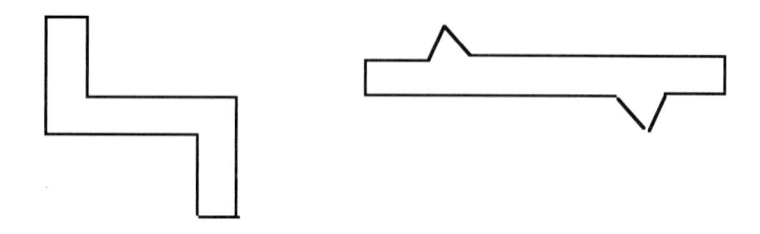

    c. A regular pentagon is one example of a figure with five rotation symmetries and five lines of symmetry.

13. a. For a right regular octagonal prism there are eight vertical planes of symmetry and one horizontal plane of symmetry, for a total of nine planes of symmetry. Four of the vertical planes go through opposite vertices and four go through midpoints of opposite sides of the bases. The horizontal plane of symmetry cuts midway through the other faces of the prism.

    b. A right cone (see the figure in #9e) has an infinite number of planes of symmetry. Any vertical plane going through the vertex of the cone is a plane of symmetry.

    c. If it is a right pyramid with a regular pentagon as base (see the figure in #9a) then there are five planes of symmetry. Each vertical plane going through a vertex of the base and the midpoint of the opposite side is a plane of symmetry.
    *[Would an oblique pentagonal pyramid also have five planes of symmetry?]*

14. For symmetry about line m, copy the segments below line m so that they form a mirror image. For symmetry about line n, copy the segments to the left and right of line n so that they form mirror images. The completed figure is shown below.

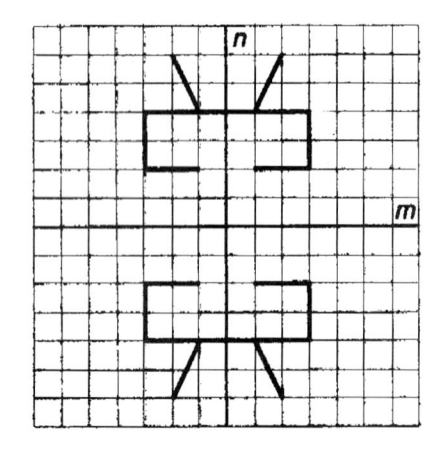

15. a. A rectangle has two lines of symmetry through midpoints of opposite sides. It has two rotation symmetries, through 180° and 360°.

   b. A regular heptagon (7 sides) has 7 lines of symmetry, one joining each vertex to the midpoint of the opposite side. It also has 7 rotation symmetries.

   c. An equilateral triangle has 3 lines of symmetry and 3 rotation symmetries.

   d. A parallelogram has no lines of symmetry unless it is also a rectangle. It does have two rotation symmetries.

16. Call the seven points A, B, C, D, E, F, and G. Point A can connect to each of the other six points to form 6 segments. Point B can connect to each of the five remaining points to form 5 more segments, point C connects to D, E, F, and G to form 4 more, etc. There are a total of 6 + 5 + 4 + 3 + 2 + 1 = 21 segments possible.

17. A regular polygon with 40 sides can be subdivided into 38 triangles. The sum of the angles in these 38 triangles is the same as the sum of the vertex angles in the regular 40-gon. Since each triangle has angles totaling 180°, the sum of the vertex angles is 180 × 38 = 6840°. There are 40 equal vertex angles in the 40-gon, so each has measure 6840 ÷ 40 = 171°.

18. a. We create the maximum number of regions if we do not let more than two lines to intersect at any point and if we make sure that all of the chords intersect each other. Then the maximum number of regions for four lines is 11. See the figure below.

   b. Here are the results so far: 1 line = 2 regions; 2 lines = 4 regions; 3 lines = 7 regions; 4 lines = 11 regions. In the sequence 2,4,7,11 the difference between successive terms is increasing by one each time. The next number in this sequence should be 11+ 5 = 16. Then the sequence is 2, 4, 7, 11, 16, 22, 29, 37, 46, 56. For 10 lines we get 56 regions.

19. a. There are three lines of symmetry and three rotation symmetries.

   b. There are four lines of symmetry and four rotation symmetries.

   c. There are two lines of symmetry and two rotation symmetries. Diagonal lines are not lines of symmetry in this figure because the larger arms would not reflect onto their opposite counterparts.

   d. There are four lines of symmetry and four rotation symmetries.

# Chapter 10 Measurement
## Section 10.1

1. a. One mile is approximately $\frac{5}{3}$ km. so 55 miles ≈ 55 × $\frac{5}{3}$ km. ≈ 92 km.
   So 55 mph ≈ 92 kph. Another common conversion approximation is 1 mile ≈ 1.6 km. Using this conversion factor we get 55 mph ≈ 88 kph.

   b. $24 \times \frac{5}{3}$ = 40 so 24 miles is approximately equal to 40 km. (Or, using 1.6 it is 38.4 km.)

3. a. The length of the scissors shown is approximately the length of $2\frac{1}{2}$ of the paper clips.

   b. The length of the scissors shown is approximately the length of 5 of the erasers.

5. a. One kilogram is 1000 grams, so 202.5 kg = 202,500 grams. We multiply kg by 1000 to get grams.

   b. There are 1000 milligrams in each gram, so the weight in milligrams is 202,500 × 1000 = 202,500,000 mg.

   c. Each kilogram is approximately 2.2 pounds. Since 202.5 × 2.2 = 445.5, the weight is approximately 446 pounds.

7. a. The thickness of a pencil lead is approximately $\frac{1}{16}$ inch.

   b. The width of a pencil is approximately $\frac{1}{4}$ inch.

   c. The diameter of a dime is approximately $\frac{11}{16}$ inch.

   d. The length of a dollar bill is approximately $6\frac{3}{16}$ inches.

   e. The width of a sheet of paper is $8\frac{1}{2}$ inches.

9. a. There are 2 pints in 1 quart and 4 quarts in 1 gallon, so 1 gallon = 8 pints.

   b. There are 2 cups per pint, so 1 gallon = 16 cups.

   c. There are 2 cups per pint, so $\frac{1}{2}$ pint = 1 cup.

11. a. There are 3 feet in 1 yard, so to convert feet to yards we divide by 3.
      12.6 ÷ 3 = 4.2        12.6 ft = 4.2 yd

    b. There are 16 ounces in 1 pound, so to convert ounces to pounds we divide by 16.
       56 ÷ 16 = 3.5        56 oz = 3.5 lb

    c. There are 16 cups in one gallon, so to convert cups to gallons we divide by 16.
       40 ÷ 16 = 2.5        40 cups = 2.5 gal

13. a. There are 100 cm in one meter, so 200 cm = 2 meters. One meter is about 1 yard or 3 feet in length. Of those given, the most realistic measure for length of a ski is 200 cm.

    b. Since 1 kg ≈ 2.2 lb, 75 kg ≈ 165 pounds. This is a realistic weight for a person. The other weights given are both less than 1 pound.

    c. A liter is close to a quart, so 48 L ≈ 12 gallons. This is a realistic automobile gas tank size. 48 mL is 1000 times smaller and 48 kL is 1000 times larger, both unrealistic.

15. a. Each person's cubit will vary. Unless your arms are fairly long your cubit is probably a bit less than 52.5 cm.

    b. Since 1 meter = 100 cm, 52.5 cm = .525 meters. To convert cubits to meters we multiply cubits by .525. Since 1 meter ≈ 3.28 feet we can convert meters to feet by multiplying meters by 3.28. Answers in the table below were rounded to the nearest whole number.

    |         | Cubits | Meters | Feet |
    |---------|--------|--------|------|
    | Length  | 300    | 158    | 517  |
    | Breadth | 50     | 26     | 86   |
    | Height  | 30     | 16     | 52   |

17. Here are the completed clues. See the answer in the back of the text for the filled in puzzle.

    *Across*
    1. 16.5 cm = <u>165</u> mm
    4. 3.15 L of water weighs approximately <u>3150</u> g
    5. .12 L = <u>120</u> mL
    8. <u>92</u> kg = 92,000 g
    9. <u>7920</u> mL of water weighs approximately 7.920 kg
    10. <u>5550</u> m = 5.55 km

    *Down*
    2. 632,000 L = <u>632</u> kL
    3. 4.5 km = <u>4500</u> m
    6. <u>432</u> g is the approximate weight of 432 mL of water
    7. <u>190</u> kg is the approximate weight of 190 L of water
    8. <u>900</u> mg = .9 g
    9. .75 m = <u>75</u> cm

19. a. If each egg weighs about 35 grams, then 12 eggs weigh about 35 × 12 = 420 g. With the carton the weight is about 450 g or .45 kg.

   b. An adult's walking step is about 2 to 3 feet, so the hose length is about 60 to 90 feet.

   c. Without taking the perspective of the drawing into account the room appears about as tall as the man. But since the man is in the foreground, he is probably not quite as tall as the door. If the door is about 7 feet tall, then the room is about 8 feet tall.

21. a. The magnified strand of hair in the picture is about 17 mm thick.

   b. Since the hair has been magnified 200 times its actual thickness is about $\frac{17}{200}$ = .085 mm.

   c. There are 1000 microns in one millimeter, so the hair is .085 × 1000 = 85 μ thick.

   d. This human hair is 85 μ thick, so it is 85 times thicker than a microcircuit wire with a thickness of 1 μ.

23. a. If it is measured to the nearest kg then it can be off by at most .5 kg either way. The weight of the washing machine is between 111.5 and 112.5 kg.

   b. Measured to the nearest tenth of a degree means that the temperature can be off by at most one half of a tenth, or .05 degrees. The range is from 38.15°C to 38.25°C.

   c. The width of the speaker is between 48.25 cm and 48.35 cm.

   d. Half of a hundredth of a kilogram is .005 kg.
   So the weight is in the range from 3.455 kg to 3.465 kg.
   (This could also be stated as 3455 g to 3465 g).

25. We can find the total weight in kg by first converting those items listed in grams to their weights in kg and then adding the weights. 1 kg = 1000 g, so we can convert grams to kg by dividing by 1000, or moving the decimal three places to the left. The tomatoes are .754 kg, the soup is .772 kg, the raisins are .425 kg, and the baking powder is .218 kg. The total weight in kilograms is .754 + .772 + 3.45 + 4.62 + .425 + 1.361 + .218 = 11.6 kg.

27. a. Rounding each amount of gasoline to the nearest 10 L and adding we get
   40 + 30 + 50 + 40 + 30 = 190 L. Since more of these were rounded down than up, an approximation of 200 L will also be close (maybe closer). 100 L at 32 cents per liter costs $32, so 200 L costs $64. (Other approximation techniques can also be used.)

   b. The exact amount of gasoline purchased was 38.2+26.8+54.3+44.7+34 = 198 L.
   The total cost is 198 × $.32 = $63.36

29. Each kg is 1000 grams, so a 24 kg bag of bird seed contains 24,000 grams of seed. At a rate of 75 g of seed used per day the seed will last 24,000 ÷ 75 = 320 days.
Since we know the number of days that the bag will last we can compute the cost per day by dividing the total cost by the number of days. $16.88 ÷ 320 = .05275. To the nearest cent this is 5 cents per day to feed the birds. (chicken feed!)

31. a. Two cubic centimeters of the garamycin weighs 80 mg, so each cubic centimeter weighs 40 mg. To get 12 injections of 24 mg each we need 12 × 24 = 288 mg of garamycin. So we need 288 ÷ 40 = 7.2 cubic centimeters of garamycin for 12 injections.

　b. The garamycin in each vial weighs 80 mg. Each injection of 60 mg will use $\frac{60}{80} = \frac{3}{4}$ of a vial. To find the number of $\frac{3}{4}$ vials in 24 vials we can divide.
$24 ÷ \frac{3}{4} = 24 × \frac{4}{3} = 32$ injections.

33. a. Light travels 299,792,458 meters in one second.

　b. The speed of light in meters per second is 299,792,458 m/sec.
There are 1000 meters in one km, so the number for the speed in km/sec will be 1000 times less. The speed is 299,792.458 km/sec.

　c. The range given by 299,792.4 plus or minus .11 is from 299,792.29 to 299,792.51
The speed used by the Conference, 299,792.458 km/sec is within this range.

35. We can either first convert kilometers to centimeters and then divide, or first find the space for each person in km and then convert this to centimeters. There are 1000 meters in a kilometer and 100 centimeters in a meter, so there are 100,000 centimeters per kilometer. <u>Method 1:</u> We convert 40,077 km to 4,007,700,000 cm and then divide 4,007,700,000 ÷ 281,000,000 = 4,007.7 ÷ 281 ≈ 14 cm per person. <u>Method 2:</u> If we space 281,000,000 people around the equator they each get 40,077 ÷ 281,000,000 ≈ .00014 kilometers of space. Multiplying by 100,000 this is 14 centimeters of space per person.

## **Section 10.2**

1. a. The width of 2 meters is greater than the width of the outstretched arms of the average adult. The height of 4 meters is about double the height of an average room.

　b. If each rectangle is 2 m wide, then the front face is 53 × 2 = 106 m wide.

　c. The height of the front face is 10 × 4 = 40 m.

　d. The area of the front face is the product of its width and height.
40 m × 106 m = 4240 m².

3. a. The trapezoid shape contains $1\frac{1}{2}$ of the geoboard units of area. If we use this trapezoid region as our unit of area we find that the octagon contains the same amount of area as 8 of these trapezoids. Its area is 8 units. This can be seen be marking off trapezoidal areas in the octagon.

   b. Using the double square as a unit of area we see that the octagon has an area of 6 units.

5. About 4 of the plastic fasteners fill the figure, so it has an area of about 4 of these units.

7. a. One square mile is a square region whose dimensions are one mile by one mile. There are 5280 feet in a length of one mile, so a square mile has dimensions of 5280 feet by 5280 feet. There are 5280 × 5280 = 27,878,400 square feet in a square mile.

   b. One acre is 43,560 square feet, so there are 27,878,400 ÷ 43,560 = 640 acres in a sq. mi.

   c. 1049 × 640 = 671,360. There are 671,360 acres in the state of Rhode Island.

9. a. Counting the sides of squares on the outside of the figure we see a perimeter of 10 units.

   b. Counting the sides of squares on the outside of the figure we see a perimeter of 8 units.

11. a. One are is an area of 10 m by 10 m. or 100 m$^2$.
    One square kilometer is 1000 m by 1000 m or 1,000,000 m$^2$.
    There are 1,000,000 ÷ 100 = 10,000 ares in 1 km$^2$.

    b. One hectare is 100 ares, so there are 10,000 ÷ 100 = 100 hectares per km$^2$.

13. a. To compute the perimeter of the rectangle we add the lengths of the sides. There are two sides with length 55 mm and two sides with length 25 mm. 55 × 2 + 25 × 2 = 160 mm. To compute the area in mm$^2$ we multiply the length by the width. The product is 55 × 25 = 1375 mm$^2$. Since there are 10 mm in 1 cm, there are 10 × 10 = 100 mm$^2$ in one cm$^2$. So the area of the rectangle is 1375 mm$^2$ = 13.75 cm$^2$.

    b. The perimeter of the parallelogram is 60 × 2 + 31.4 × 2 = 182.8 ≈ 183 mm.
    The area is the product of the base and the height. 60 × 30 = 1800 mm$^2$ = 18 cm$^2$.

15. a. The circumference of the circle is π times the diameter. 54.7 × π ≈ 172 mm.
    The area of the circle is π times the square of the radius. Since the diameter is 54.7 mm the radius is 54.7 ÷ 2 = 27.35 mm. 3.1416 × 27.35$^2$ ≈ 2350 mm$^2$ = 23.5 cm$^2$.

    b. The perimeter of the trapezoid is the sum of the sides. 39+33.6+36+48.6 ≈ 157 mm.
    The area is the product of the height times the average of the bases. The bases are the two parallel sides. The average of their lengths is (48.6+33.6)÷2 = 41.1 mm.
    The area is 41.1 × 36 ≈ 1480 mm$^2$ = 14.8 cm$^2$.

17. a. To find the perimeter we need to know the length of the third side. Call this length x. We can use the Pythagorean Theorem to find x, but first we need to know how much of the 96 mm side is to the left of the 32 mm altitude and how much is to the right. We also use the Pythagorean Theorem to answer this question. If we call the length to the left of the altitude l, then $l^2 + 32^2 = 40^2$. Solving for l we get that l = 24, so the part of the 96 mm base to the right of the altitude is 96 − 24 = 72 mm. Now we have $72^2 + 32^2 = x^2$. Solving for x gives x ≈ 79 mm, so the perimeter of the triangle is 79 + 96 + 40 = 215 mm. To find the area we will use the 96 mm side as the base. The area of a triangle is equal to $\frac{1}{2} \times b \times h$, so for this triangle A = $\frac{1}{2} \times 96 \times 32$ = 1536 mm$^2$ = 15.36 cm$^2$.

   b. Again we are missing the length of one side of the trapezoid and again we will use the Pythagorean Theorem to find it. Form a right triangle in the top of this trapezoid by drawing a 30 mm altitude from the upper left vertex to the 43.4 mm base. Call the length of the unknown side of the trapezoid x. The right triangle we have just constructed has legs of length 30 mm and 43.4 − 27.4 = 16mm. The length of the hypotenuse is x. By the Pythagorean Theorem $30^2 + 16^2 = x^2$. Solving we get x = 34. The perimeter of the trapezoid is 34 + 43.4 + 30 + 27.4 ≈ 135 mm. The area is the product of the height times the average of the bases. The bases are the two parallel sides. The average of their lengths is (27.4+43.4)÷2 = 35.4 mm. The area is 35.4 × 30 = 1062 mm$^2$ = 10.62 cm$^2$.

19. Subdivide this polygon into one large triangle on top and one large rectangle between two small triangles on the bottom. The large triangle on top has a base of 4 + 1 + 1 = 6 cm and height of 2 cm, so its area is $\frac{1}{2} \times 6 \times 2 = 6$ cm$^2$. The area of the rectangle is 8 cm$^2$. The area of the small triangle to the left is $\frac{1}{2} \times 1 \times 1 = \frac{1}{2}$ cm$^2$ and the area of the small triangle to the right is $\frac{1}{2} \times 1 \times 2 = 1$ cm$^2$. The total area is $6 + 8 + \frac{1}{2} + 1 = 15\frac{1}{2}$ cm$^2$.

21. a. Adding the bases and dividing by 2 we get (10 + 18) ÷ 2 = 14. The height is 10 cm because it is the perpendicular distance between the bases. The area of the trapezoid is 14 × 10 = 140 cm$^2$.

   b. We can find the area of the shaded region by finding the area of the original rectangle and then subtract the areas of the circle and square that have been removed. The circle has an area of $\pi \times 5^2 \approx 78.54$ cm$^2$. Since the piece cut out of the upper right is a square, both of its dimensions are 6 and its area is 36 cm$^2$. The large rectangle has a length of 22.3 cm and a width of 8 + 6 = 14 cm. The area of the rectangle is 22.3 × 14 = 312.2 cm$^2$. The area of the shaded region is 312.2 − 78.54 − 36 = 197.66 ≈ 198 cm$^2$.

23. The circumference of a circle is found by multiplying π times the diameter (or by multiplying π times twice the radius). The circumference of the first circle is 2π ≈ 6.3 m. The circumference of the second circle is 4π ≈ 12.6 m. The circumference of the third circle is 8π ≈ 25.1 m. The circumference of the fourth circle is 16π ≈ 50.3 m. When the radius of a circle is doubled the circumference is also doubled.

25. The starfish approximately fits in a 10 by 10 square on the grid, so if we approximate the area not covered by the starfish within this 100 square cm and then subtract from 100 we will have an approximation for the area covered by the starfish. There are about 21-25 squares in the area above the two arms that reach across the 10 by 10 square. In the lower right are about 13-15 squares, and in the lower left are about 12-14 squares. So there are about 46-54 square centimeters not covered by the starfish inside the 100 square centimeter grid. There is also about 0-1 cm$^2$ of starfish outside the 10 by 10 square. This gives an area covered by the starfish of between 100 – 54 = 46 cm$^2$ and 101 – 46 = 55 cm$^2$. (Different techniques can be used for approximating here, and answers will vary somewhat.)

27. a. The perimeter of the square is 30 × 4 = 120 mm.
    The circumference of the circle is 38.2 × $\pi$ ≈ 120 mm.

    b. The square's area is 30 mm × 30 mm = 900 mm$^2$.
    The area of the circle is (19.1)$^2$ × $\pi$ ≈ 1146 mm$^2$, which is about 246 mm$^2$ greater.

29. a. First we will find the number of dots on each page. There are 50 × 50 = 2500 dots per square inch of printed surface. Each page has 8 × 10 = 80 square inches of printed surface, so each page contains 2500 × 80 = 200,000 dots. Each directory contains 3400 pages, so each directory contains 3400 × 200,000 = 680,000,000 dots.

    b. We need to find the number of directories required to have a total of 6,000,000,000 dots. We divide 6,000,000,000 ÷ 680,000,000, which is equivalent to 6000 ÷ 680 ≈ 8.8. (It would also be appropriate to make this calculation using scientific notation). About 8.8 directories would be required to contain one dot for each person in the world.

31. The area of each roll of paper in Type A is 75 × 150 = 11250 cm$^2$. The package of four rolls contains 4 × 11250 = 45,000 cm$^2$. Type A costs $2.99.
    Type B has a total area of 88 × 500 = 44,000 cm$^2$. Type B costs $3.19. With Type A you get more paper for a lower cost, so it is a better deal.

33. The glass is sold per square meter. We can either convert the measurements from cm to meters and then find the area or we can find the area in cm$^2$ and then convert the areas to square meters. Each meter is 100 cm, so there are 100 × 100 = 10,000 cm$^2$ in a square meter. Using cm, we have an area of 58 × 30 = 1740 cm$^2$ for the glass in the first frame and an area of 40 × 60 = 2400 cm$^2$ for the glass in the second frame. The total area of glass is 1740 + 2400 = 4140 cm$^2$ = .414 m$^2$. At a cost of $20 per square meter the glass for the two frames will cost $20 × .414 = $8.28.

35. A 50-kg bag of fertilizer contains 50,000 grams of fertilizer, since there are 1000 grams in a kilogram. At a rate of 35 grams per square meter of lawn, 50,000 ÷ 35 ≈ 1430 square meters of lawn can be fertilized.

37. a. One way to find the area to be covered with wallpaper is to find the area of the whole wall and subtract the areas of the window and the fireplace. The area of the wall is 540 × 240 = 129,600 cm$^2$. The area of the window is 105 × 210 = 22,050 cm$^2$.
    The area of the fireplace is 125 × 135 = 16,875 cm$^2$.
    The area to be wallpapered is 129,600 – 22,050 – 16,875 = 90,675 cm$^2$.

b. A roll of wallpaper that is 12.8 meters by 53 cm is also 1280 cm by 53 cm. Its area is 1280 × 53 = 67,840 cm². More than one roll is needed. Two rolls should be enough.

39. a. We add the areas of the two rectangles. 8.7 × 11 = 95.7 m² and 8 × 9.5 = 76 m². The area of the base of the house is 95.7 m² + 76 m² = 171.7 m². The value for tax assessment is $389 × 171.7 = $66,791.30.

   b. If the tax rate is $73 for every $1000 of assessed value then for this house the tax is $73 × 66.7913 ≈ $4875.76.

41. For the trundle wheel to measure off one meter along the ground each time it makes one revolution it needs to have a circumference of 1 meter. We need to know the diameter of the wheel to find the amount of plywood needed to make the wheel. If the circumference is one meter, then the diameter of the circle is 1 ÷ π, since C = πd. 1 ÷ π ≈ .318, so the diameter of the circle is about .318 m or 31.8 cm. A square piece of plywood needs to be at least 31.8 cm by 31.8 cm in order to cut the trundle wheel.

43. a. The diameter of the building is 35.36 m, so this is also the diameter of a floor of the building. The floors are circles, so the area of a floor is the area of a circle with diameter 35.36 m. A = π × r² = π × (17.68)² ≈ 982 square meters.

   b. We need to find the length of the side of a square whose area is 982 and then multiply by four to find its perimeter. We use the square root. $\sqrt{982}$ ≈ 31.3 and 31.3 × 4 ≈ 125. So a square with area 982 m² would have a perimeter of about 125 m.

   c. The diameter of the Peachtree Plaza Hotel is 35.36 m, so its perimeter (circumference) is π × 35.36 ≈ 111 m.

   d. The perimeter of the square is 125 – 111 = 14 m longer than the perimeter of the hotel.

   e. If the hotel had a square base rather than a circular base then it would have required an extra 230 × 14 = 3220 m² of wall area in order to enclose the same space.

45. First we will find the distance along the ground that one wheel needs to travel in order to turn the contraption in one complete circle. The distance between opposite pairs of wheels is 300 cm, so the circle created by their track has a diameter of 300 cm. Then its circumference is π × 300 ≈ 942 cm. This is the distance that a wheel needs to travel in order for the contraption to make a complete circle. Each wheel has a diameter of 75 cm, so each time it makes a revolution it has traveled π × 75 ≈ 236 cm. The number of revolutions each wheel will make in traveling 942 cm is 942 ÷ 236 ≈ 4 revolutions.

204　　　　　　　　　　　　*Chapter 10 Measurement*

## Section 10.3

1. a. The wheel is a rectangular prism with a cylinder removed from the center. We can find the volume of the wheel by subtracting the volume of the cylinder from the volume of the rectangular prism. The dimensions of the rectangular prism are 1 meter by 1 meter by 20 centimeters, or, converting to centimeters, 100 cm by 100 cm by 20 cm. Its volume is $100 \times 100 \times 20 = 200{,}000$ cm$^3$. The cylinder that was removed has diameter 46 cm, so its radius is 23 cm. Its height is 20 cm, the thickness of the wheel. The volume of the cylinder is the area of the circular base times the height, $\pi \times 23^2 \times 20 \approx 33{,}238$ cm$^3$. The volume of the wheel is 200,000 cm$^3$ – 33,238 cm$^3$ = 166,762 cm$^3$.

   b. If the wheel weighs 7 grams per cubic centimeter its weight in grams is $166{,}762 \times 7 = 1{,}167{,}334$ grams. There are 1000 grams in a kilogram, so the weight in kilograms is about 1167 kg.

3. a. <u>Using unit (i):</u> The cube has dimensions 3 by 3 by 3. The volume is 27 cubic units. The surface area consists of 6 faces each made up of 9 square units, so the total surface area is $6 \times 9 = 54$ square units.
   <u>Using unit (ii):</u> Unit (ii) has a volume of 8 small cubes. The volume of the cube in 3a is $27 \div 8 = 3\frac{3}{8}$ cubic units. One surface of unit (ii) has an area of 4 small squares. Since the cube in 3a has a surface area of 54 small squares, its surface area in terms of unit (ii) is $54 \div 4 = 13\frac{1}{2}$ square units.

   b. <u>Using unit (i):</u> The figure contains 12 cubes on the bottom level, 9 on the next level, 6 on the next level, and 3 on the top level. The volume is $12 + 9 + 6 + 3 = 30$ cubic units. The front and back surfaces each have area $1 + 2 + 3 + 4 = 10$ square units, the left side has an area of 12 square units, the bottom has an area of 12 square units, and the fronts and tops of the steps have a total surface area of $3 \times 8 = 24$ square units. The total surface area of the figure is $10 + 10 + 12 + 12 + 24 = 68$ square units.

   <u>Using unit (ii):</u> Unit (ii) has a volume of 8 small cubes. The volume of the figure in 3b. is $30 \div 8 = 3\frac{3}{4}$ cubic units. One surface of unit (ii) has an area of 4 small squares. Since the cube in 3a has a surface area of 68 small squares, its surface area in terms of unit (ii) is $68 \div 4 = 17$ square units.

5. a. There are 36 inches in one yard. One cubic yard has dimensions 36 inches by 36 inches by 36 inches, so there are $36^3 = 46{,}656$ cubic inches in 1 yd$^3$.

   b. There are 1000 mm in one meter. One cubic meter has dimensions 1000 mm by 1000 mm by 1000 mm, so there are $1000^3 = 1{,}000{,}000{,}000$ cubic millimeters in 1 m$^3$.

7. a. <u>Volume:</u> The volume of the square pyramid is one third of the volume of a rectangular prism with the same base. The volume formula is $V = \frac{1}{3} \times B \times h$, where B is the area of the base and h is the altitude. For this pyramid the base is a square with area $6 \times 6 = 36$ cm$^2$. The altitude is 4 cm (5 cm is the slant height and is not needed for finding the volume). The pyramid's volume is $\frac{1}{3} \times 36 \times 4 = 48$ cm$^3$.
<u>Surface area:</u> The surface consists of the four triangular sides and the square base. The area of the base is 36 cm$^2$. The area of each triangular face is $\frac{1}{2} \times 6 \times 5 = 15$ cm$^2$. The total surface area is $4 \times 15 + 36 = 96$ cm$^2$.

b. <u>Volume:</u> This is a triangular prism. Its volume is the product of the area of a triangular base (side) times the altitude. For this prism the altitude is the distance between the triangular bases which is 10 cm. The area of a base is $\frac{1}{2} \times 4 \times 3 = 6$ cm$^2$.
The volume is $6 \times 10 = 60$ cm$^3$.
<u>Surface area:</u> The surface consists of the two triangular faces and three rectangular faces. The area of each triangle is 6 cm$^2$. The area of the bottom rectangle is $4 \times 10 = 40$ cm$^2$. The other two rectangles also have length 10, but we need to use the Pythagorean theorem to find their other dimension. Using the triangle formed by half of one triangular face, if d represents the length of the missing dimension, then $2^2 + 3^2 = d^2$. So $d^2 = 13$ and $d = \sqrt{13} \approx 3.6$. Then the other two rectangular faces each have areas approximately $3.6 \times 10 = 36$ cm$^2$. The total surface area is $6 + 6 + 40 + 36 + 36 = 124$ cm$^2$.

9. a. <u>Volume:</u> To compute the volume of a sphere we only need to know its radius. The formula is $V = \frac{4}{3} \times \pi \times r^3$. The cube of a radius of 3 cm is 27 cm$^3$.
The volume of the sphere is $\frac{4}{3} \times \pi \times 27 \approx 113$ cm$^3$.
<u>Surface area:</u> The formula for the surface area of a sphere is $4 \times \pi \times r^2$. The surface area of a sphere with radius 3 cm is $4 \times \pi \times 3^2 \approx 113$ cm$^2$.
*[Note that even though the numbers for the volume and surface area are the same here, the units are different. Will the volume and surface area always have the same numbers for any sphere? Why were they the same this time?]*

b. <u>Volume:</u> To compute the volume of a cylinder, multiply the area of the circular base by the height. The base of this cylinder has a diameter of 5 cm, so its radius is 2.5 cm. Its area is $\pi \times 2.5^2 \approx 19.6$ cm$^2$. The height is 10 cm, so the volume is $19.6 \times 10 = 196$ cm$^3$.
<u>Surface area:</u> The cylinder has three surfaces, the two circular bases and the side of the cylinder. If you imagine the side as a label on a can you can see that it can be unrolled to form a rectangle. One dimension of the rectangle is 10 cm and the other dimension is the circumference of the circle. The circumference is $\pi \times 5 \approx 15.7$ cm, so the area of the rectangle is $15.7 \times 10 = 157$ cm$^2$. The area of each circular base is 19.6 cm$^2$.
The total surface area is $157 + 19.6 + 19.6 \approx 196$ cm$^2$.

11. a. Volume: To compute the volume of the trapezoidal prism we multiply the area of the trapezoidal base by the height. The area of a trapezoid is found by taking half of the sum of the two parallel sides and multiplying by the altitude. For this trapezoid we get an area of $(4+9) \times \frac{1}{2} \times 6 = 13 \times 3 = 39$ cm$^2$. So the volume is $39 \times 15.4 \approx 601$ cm$^3$.

Surface area: There are 6 surfaces, two trapezoids and four rectangles. We have found the area of each trapezoid to be 39 cm$^2$. Each of the four rectangular faces has 15.4 cm for one of its dimensions. For each rectangle the other dimension is one of the four sides of the trapezoid. The four areas are: $9 \times 15.4 = 138.6$ cm$^2$, $4 \times 15.4 = 61.6$ cm$^2$, $6.5 \times 15.4 = 100.1$ cm$^2$, and $6.5 \times 15.4 = 100.1$ cm$^2$.
The total surface area is $39 + 39 + 138.6 + 61.6 + 100.1 + 100.1 \approx 478$ cm$^2$.

b. Volume: The volume of the pyramid is one third the area of the base times the height. The base is an equilateral triangle with each side having length 8 cm. In order to find its area we need to know its altitude, the perpendicular distance from a vertex to the opposite side. In the case of an equilateral triangle the altitude also bisects the opposite side. We use the Pythagorean theorem to find the length of the altitude. In the figure below this is represented by n. Then $4^2 + n^2 = 8^2$. So $n^2 = 8^2 - 4^2 = 64 - 16 = 48$, and $n = \sqrt{48} \approx 6.9$

So the area of the triangular base is $\frac{1}{2} \times 8 \times 6.9 \approx 27.6$ cm$^2$.

The volume of the pyramid is $\frac{1}{3} \times 27.6 \times 15.7 \approx 144$ cm$^3$.

Surface area: There are 4 surfaces. One is the triangular base whose area we just found to be 27.6 cm$^2$. The three lateral faces are each triangles with bases of 8 cm and heights of 15.9 cm. The area of one of these triangles is $\frac{1}{2} \times 8 \times 15.9 = 63.6$ cm$^2$. The total surface area is $3 \times 63.6 + 27.6 \approx 218$ cm$^2$.

13. a. The volume of a cone is one third of the volume of the cylinder with the same base and height. In this case the radius of the base is 1.5 cm, so the area of the base is $\pi \times 1.5^2 \approx 7.1$ cm$^2$. The volume of the cone is $V = \frac{1}{3} \times 7.1 \times 4 \approx 9.5$ cm$^3$.

b. The volume of a prism is the area of the base multiplied by the height. It does not matter whether the prism is right or oblique. It is important however when the prism is oblique to be sure that the height is the perpendicular distance between the bases. For this prism the volume is $4 \times 1 \times 3 = 12$ cm$^3$.

15. a. The volume of the tank is 50 × 25 × 30 = 37,500 cm³. There are 1000 cm³ in a liter, so the tank will hold 37,500 ÷ 1000 = 37.5 L of water.

    b. Each fish would have 37,500 ÷ 30 = 1250 cm³ of space.

    c. 37,500 ÷ 3000 = 12.5, so 12 goldfish could live in this tank.

    d. The bottom of the tank needs 50 × 25 = 1250 cm² of glass. The front and back need 50 × 30 = 1500 cm² of glass each. The two sides need 30 × 25 = 750 cm² of glass each. A total of 1250 + 3000 + 1500 = 5750 cm² of glass is needed for the tank.

17. We need to find the total volume of the five rooms in the house. The first room has a volume of 4 × 5 × 2.4 = 48 m³, the second room has volume 4 × 4 × 2.4 = 38.4 m³, the third room has volume 6 × 4 × 2.4 = 57.6 m³, the fourth room has volume 6 × 6 × 2.4 = 86.4 m³, and the fifth room has volume 6 × 5.5 × 2.4 = 79.2 m³. The total volume of the rooms is 309.6 m³. A 21,000 Btu unit will be adequate.

19. The volume of type A is 60 × 60 × 150 = 540,000 cm³. Its cost is $339, so the cubic centimeters per dollar are 540,000 ÷ 339 ≈ 1593 cm³. For type B the volume is 55 × 72 × 160 = 633,600 cm³. The cm³ per dollar are 633,600 ÷ 379 ≈ 1672 cm³. Type B gives more volume per dollar.

21. We need to find the lateral surface area of the cylindrical columns. We will not be painting the circular bases of the columns, only the sides. If the surface of the side of the cylinder were unrolled it would be a rectangle with height 22 feet and length equal to the circumference of the base of the column. The diameter is 2.5 feet, so the circumference is π × 2.5 ≈ 7.85 ft. Each column has 7.85 × 22 ≈ 173 square feet of area to paint. To paint all 20 columns enough paint must be purchased to cover 173 × 20 = 3460 square feet. Each gallon covers 350 square feet. Since 3460 ÷ 350 ≈ 9.9, 10 gallons will be needed.

23. a. To find the volume of each pyramid we use the formula $V = \frac{1}{3} \times B \times h$, where B is the area of the base and h is the height of the pyramid. We are given the perimeter of the base for each pyramid, but not its area. Both have square bases, so the length of a side is 1/4 of the perimeter and the area is the square of this number. For the Great Pyramid the area of the base is $(930 \div 4)^2 \approx 54,056$ m². The volume of the Great Pyramid is $V = \frac{1}{3} \times 54056 \times 148 \approx 2,667,000$ m³. For the Transamerica Pyramid, the area of the base is $(140 \div 4)^2 = 1225$ m², and the volume is $V = \frac{1}{3} \times 1225 \times 260 \approx 106,000$ m³. The volume of the Great Pyramid is about 2,667,000 ÷ 106,000 ≈ 25 times larger.

b. We are comparing the total areas of the four walls of each pyramid (not including the square floors). Each wall is a triangle. For the Great Pyramid each wall has a base of 232 m and an altitude (slant height) of 188 m, so the area of each wall is $\frac{1}{2} \times 232.5 \times 188 = 21{,}855$ m². The total area of the walls is $21{,}855 \times 4 = 87{,}420$ m². For the Transamerica Pyramid the area is $\frac{1}{2} \times 35 \times 261 \times 4 = 18{,}270$ m². The surface area of the Great Pyramid is about $87{,}420 \div 18{,}270 \approx 5$ times larger.

25. a. Each silo is a cylinder whose base has area $\pi \times 3^2 \approx 28.27$ m². The volume of the cylinder is $28.27 \times 18 \approx 509$ m³.

  b. If a blower can load 1 m³ in 3 minutes then it can load 20 m³ in $3 \times 20 = 60$ minutes. At a rate of 20 m³ per hour it will take $509 \div 20 \approx 25.5$ hours to fill one silo.

27. a. The surface area of the cylinder includes the areas of the two circular bases and the area of the side of the cylinder. One dimension of the side is the circumference of the base and the other dimension is the height. The area of the side is $\pi \times 1 \times 2 \approx 6.28$ m². The area of each base is $\pi \times .5^2 \approx .785$ m². The total area is $6.28 + .785 + .785 = 7.85$ m².

  b. If each square meter of metal weighs 92 kg., then the cylinder weighs approximately $92 \times 7.85 = 722.2$ kg.

29. There are several ways to approach this problem. One way to find the number of cubes with no faces painted is to imagine peeling the outer layer off of the cube. The original cube has dimensions 10 by 10 by 10. With the outer layer peeled off the cube will have dimensions 8 by 8 by 8. This means that $8 \times 8 \times 8 = 512$ cubes are unpainted. Of the remaining cubes, eight (one at each vertex) have three faces painted. On each face of the large cube there are $8 \times 8 = 64$ cubes with one face painted, so there are a total of $6 \times 64 = 384$ cubes with one face painted. Along each edge there are 8 cubes with two faces painted. There are 12 edges on the large cube, so there are $12 \times 8 = 96$ cubes with two faces painted.
Checking, this is a total of $512 + 8 + 384 + 96 = 1000$ cubes.
To generalize this result it is helpful to notice that the number 8 showed up often in the calculations for a 10 by 10 by 10 cube. If we had started with a 7 by 7 by 7 cube, then the number 5 would be useful in calculating the number of cubes with 0, 1, and 2 faces painted. For a cube with dimensions n by n by n, the number n – 2 plays this role.
There are $(n-2) \times (n-2) \times (n-2) = (n-2)^3$ cubes unpainted.
There are $6 \times (n-2)^2$ cubes with one face painted.
There are $12 \times (n-2)$ cubes with two faces painted.
There are eight cubes with three faces painted.

31. a. The formula for the surface area of a sphere is $4\pi r^2$ where r is the radius. The original drop has a radius of 2 mm, so its surface area is $4\pi (2)^2 = 16\pi$. Each smaller drop has a radius of 1 mm and surface area of $4\pi (1)^2 = 4\pi$. Eight of the smaller drops will have a total surface area of $32\pi$, which is twice as much as the original drop.

b. If we further subdivide these smaller drops to eight drops of diameter 1 mm and radius .5 mm, then the total surface area of 64 of these even smaller drops will equal $64 \times 4\pi (.5)^2 = 64\pi$. This is twice again the surface area, or four times the surface area of the original drop.

c. Each time the splitting process is carried out the surface area is the double of the previous amount. If the process is carried out 3 times the surface area will be $2^3 = 8$ times greater. If the process is carried out 20 times the surface area will be $2^{20} \approx 1$ million times greater.

33. a. To find the volume of the block of cement we can find the volume of the water displaced by it. (This is the discovery that prompted Archimedes to shout "Eureka!") The volume of a cylinder with a diameter of 2 feet (radius of 1 foot = 12 inches) and a height of 1.5 inches is $\pi \times 12^2 \times 1.5 \approx 679$ cubic inches. So the volume of the cube is also approximately 679 in$^3$.

b. To find the length of an edge of the cube we need to find the number whose cube is 679. This means that we want the cube root of 679. $\sqrt[3]{679} \approx 8.8$, so the length of the edge of the cube is about 8.8 inches.

35. a. The diameter of the globe is 28 feet. Using $C = \pi d$, we get a circumference of approximately $28 \times 3.14 \approx 88$ feet.

b. The radius is 14 feet. Using S.A. = $4\pi r^2$, we get a surface area of approximately $4 \times 3.14 \times 14^2 \approx 2462$ feet.

c. Each plate is the same size, so divide the total by 28. $2462 \div 28 \approx 88$ square feet.

d. The scale factor for the globe is 1 inch = 24 miles. The 6-mile height is 1/4 of 24 miles. On the globe this would correspond to a height of 1/4 inch.

## **Chapter 10 Test**

1. a. Gram  b. Milliliter  c. Meter

   d. Kilogram  e. Square meter  f. Cubic meter

2. a. 1 ft. = 12 in., so 3.5 ft = <u>42</u> in.  b. 1 yd$^2$ = <u>9</u> ft$^2$

   c. 4 qt. = 1 gal., so 3.4 gal. = <u>13.6</u> qt.  d. 1 qt = 32 oz, so 2.5 qt = <u>80</u> oz

   e. 1 yd$^3$ = 27 ft$^3$, so 2.5 yd$^3$ = <u>67.5</u> ft$^3$  f. 1 lb = 16 oz, so 2.75 lb = <u>44</u> oz

3. a. 1.6 g = <u>1600</u> mg  b. 4.7 m = <u>470</u> cm

   c. 5.2 km = <u>5200</u> m  d. 2500 mL = <u>2.5</u> L

   e. 1.6 cm$^2$ = <u>160</u> mm$^2$  f. 1 m$^3$ = <u>1,000,000</u> cm$^3$

4.  a. 1.6 L of water weighs approximately <u>1600</u> g.

 b. 32° Fahrenheit equals <u>0</u>° Celsius.

 c. 55 cm$^3$ have a volume of <u>55</u> mL.

 d. 2 dm$^3$ of water weigh <u>2</u> kg.

 e. 1 km equals approximately <u>3/5 (or .6)</u> mi.

 f. 1 kg of water weighs approximately <u>2.2</u> lb.

5.  A measurement is generally precise to within half of the smallest unit shown in either direction.

 a. A 5.3 kg bag of dog food is between 5.25 kg and 5.35 kg.

 b. An 85 g tube of toothpaste is between 84.5 g and 85.5 g.

 c. A 4.12 oz box of cake mix is between 4.115 oz and 4.125 oz.

6.  Unit (i) has an area of half of one geoboard square. Counting squares and half squares in the figure we see that it has an area of 13 squares, or 26 units using unit (i).
 Unit (ii) has an area of two geoboard squares. Using unit (ii) as a measure of area, the figure has an area of 13 ÷ 2 = 6.5 units.

7.  a. The area of a parallelogram is the product of the base times the height. This parallelogram has area 7 × 3 = 21 cm$^2$. The perimeter is 7 + 4 + 7 + 4 = 22 cm.

 b. The area of a triangle is half the product of the base times the height. This triangle has area $\frac{1}{2}$ × 14 × 24 = 168 cm$^2$. The perimeter is 25 + 14 + 25 = 64 cm.

 c. The area of a trapezoid is the average of the bases multiplied by the height. This trapezoid has area $\frac{1}{2}$ × (4+8) × 3 = 18 cm$^2$. The perimeter is 8 + 5 + 4 + 3 = 20 cm.

 d. The area of a circle is π r$^2$. This circle has a diameter of 4 cm, so its radius is 2 cm. Its area is 4 π ≈ 12.6 cm$^2$. The circumference is the product of pi times the diameter, so it is also equal to 4 π ≈ 12.6 cm, but notice that area is measured in cm$^2$ and circumference is measured in cm, so they are not the same measurements.

8.  The cylinder formed has a circumference of 20 cm. Since C = π d, it is also true that d = C ÷ π. The diameter of the cylinder is 20 ÷ π ≈ 6.4 cm.

9. a. If we use the 9 cm side for the base of the triangle then we also need to know the length of the side that is perpendicular to it, because that is the altitude. We can use the Pythagorean theorem to do this. If we call the length of this side h, then $9^2 + h^2 = 15^2$. So $h^2 = 15^2 - 9^2 = 225 - 81 = 144$, and $h = \sqrt{144} = 12$.
Then the area of the triangle is $\frac{1}{2} \times 9 \times 12 = 54$ cm$^2$.

b. To find the area of the shaded part we can subtract the area of the unshaded isosceles triangle from the area of the whole figure.
The area of the whole figure is $\frac{1}{2} \times 8 \times (6+4) = 40$ cm$^2$.
The area of the unshaded part is $\frac{1}{2} \times 8 \times 6 = 24$ cm$^2$.
The area of the shaded part is $40 - 24 = 16$ cm$^2$.

10. The figure has 9 cubes in the bottom level, 6 in the middle level, and 3 on top. Its volume is $9 + 6 + 3 = 18$ cubic units using unit (i). Unit (ii) contains 8 small cubes, so the volume of the figure is $18 \div 8 = 2.25$ cubic units using unit (ii).
One way to count the surface area is to find the total areas of the surfaces facing up, down, forward, back, left, and right. From one viewpoint there are 9 squares facing up, 9 facing down, 6 facing forward, 6 facing back, 9 facing left, and 9 facing right. (Another viewpoint would give 6 left and 9 front, but the total will be the same). The total surface area using unit (i) is $18 + 12 + 18 = 48$ square units. A face of unit (ii) has an area of 4 squares, so the surface area using unit (ii) is $48 \div 4 = 12$ square units.

11. a. There are 3 layers of 9 cubes each, so the volume is 27 cm$^3$.
The surface area is $9 + 9 + 15 + 15 + 9 + 9 = 66$ cm$^2$.

b. There are 25 cubes on the bottom, 27 in the 3 by 3 by 3 cube in the middle, and 2 on top. The volume is $25 + 27 + 2 = 54$ cm$^3$.
There are several ways to count the surface area. Using the six different viewpoints we get areas of 25 from the top, 25 from the bottom, and 16 from each of front, back, left, and right, for a total of $50 + 64 = 114$ cm$^2$. There are other ways to count it, but they should also result in a surface area of 114 cm$^2$.

c. Counting the volume of each layer starting on the bottom, we get a total volume of $9 + 8 + 6 + 3 + 1 = 27$ cm$^3$. Looking from either top or bottom, we see 9 square units. Looking from any of the other four perspectives, we see 5+4+3 squares.
The total surface area is $9 + 9 + 4(5 + 4 + 3) = 18 + 4(12) = 18 + 48 = 66$ cm$^2$.

12. a. The area of the square base is 14 × 14 = 196 cm². 
    The volume is $\frac{1}{3}$ × 196 × 24 = 1568 cm³.

b. The area of the circular base is π × 9² ≈ 254.5 cm². The volume is approximately 254.5 × 19 = 4835.5 cm³. If the area of the base is not rounded before multiplying by 19, the approximation of the volume is slightly different, at 4834.9 cm³.

c. The area of the circular base is π × 7² ≈ 153.938 cm².
    The volume of the cone is approximately $\frac{1}{3}$ × 153.938 × 24 ≈ 1231.5 cm³.

d. The volume of the rectangular prism is 14 × 15 × 19 = 3990 cm³.

13. a. There are six surfaces on the rectangular prism. Two have area 15 × 19 = 285 cm², two have area 15 × 14 = 210 cm², and two have area 14 × 19 = 266 cm². The total surface area is 2 × 285 + 2 × 210 + 2 × 266 = 1522 cm².

b. The cylinder has two circular surfaces, each with area π × 9² ≈ 254.5 cm², and one surface that is a rectangle when it is unrolled. One dimension of this rectangle is the circumference of the circle, 18π cm, and the other is the height, 19 cm. The total surface area is 254.5 + 254.5 + 19 × 18π ≈ 1583.4 cm³. (Rounding at different points in the process of calculating can give slightly different answers).

c. The formula for the surface area of a sphere is 4πr².
    For a sphere with radius 14 cm the surface area is 4π(14)² = 784 π ≈ 2463.0 cm².

d. The square pyramid has five surfaces. Four of them are triangles with areas $\frac{1}{2}$ × 14 × 25 = 175 cm². The area of the base is 14 × 14 = 196 cm².
    The total surface area is 4 × 175 + 196 = 896 cm².

14. Measuring first in cubic feet we get 11 × 11 × .8 = 96.8 cubic feet. There are 27 cubic feet in each cubic yard, so the volume is also 96.8 ÷ 27 ≈ 3.6 cubic yards.

15. Type A has an area of 5 × .3 = 1.5 m² and has a unit cost of $3.70 ÷ 1.5 ≈ $2.47 per m².
    Type B has an area of 4 × .35 = 1.4 m² and has a unit cost of $3.50 ÷ 1.4 = $2.50 per m².
    Type A is the better buy (but not by much).

16. During the 1600 km trip the car will use 1600 ÷ 13 ≈ 123.08 L of gasoline at a cost of 123.08 × .52 = $64.00. The total cost is $414 + $64 = $478.

# Chapter 11  Motions in Geometry
## Section 11.1

1. a. $\angle B \leftrightarrow \angle T$     b. $\angle M \leftrightarrow \angle D$
   c. $\angle K \leftrightarrow \angle R$     d. $MB \leftrightarrow DT$
   e. $BK \leftrightarrow RT$     f. $MK \leftrightarrow DR$
   g. $\triangle BMK \cong \triangle TDR$

3. a. To construct a segment of length r + s, we mark an endpoint for the new segment, then set a compass open to the length of r and place one compass point on the new endpoint so that we can copy a segment of length r. Then copy a segment of length s in the same way, placing the left end of this segment on the right end of the segment of length r.

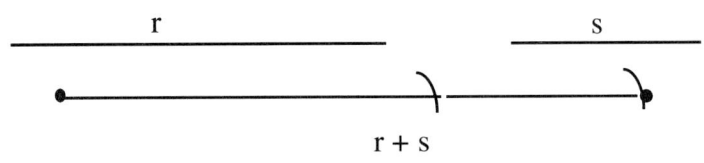

   b. To construct a segment of length r – t, copy a segment of length r as in part a, then copy a segment of length t with its right endpoint on the right endpoint of the first segment. The segment from the left endpoint of r to the left endpoint of t is of length r – t.

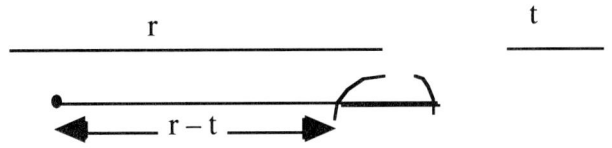

   c. To construct a segment of length r + (s – t) we construct a segment of length r as in part a, and then construct a segment of length s – t as in part b, placing the left endpoint of the segment of length s – t on the right endpoint of the segment of length r.

5. To construct an angle that is congruent to angle RST, follow the six step procedure given in Example C of section 11.1.

7. To construct the perpendicular bisector of segment AB, follow the three step procedure given in Example J of section 11.1.

9. To construct a line through K that is perpendicular to line n, follow the three step procedure given in Example L of section 11.1.

11. Extend segment RS to point B, then use a compass and locate point A so that AS = SB. Use a compass to draw arcs intersecting at point D so that D is equidistant from A and B. Then segment DS is perpendicular to segment RS.

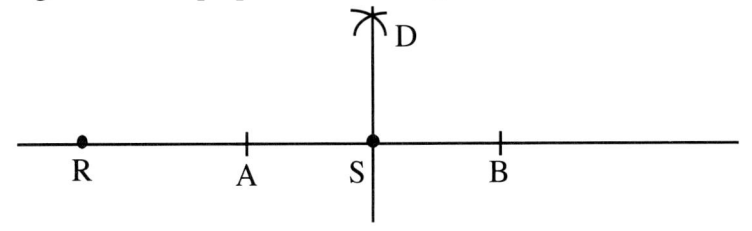

13. To use the Mira to obtain the construction in number 9, place the Mira so that one end of it is on point K and so that it crosses line n. We know that the Mira is exactly perpendicular to line n when the reflection of one side of line n exactly coincides with the other side of line n. Then we can draw the perpendicular through K using the edge of the Mira.

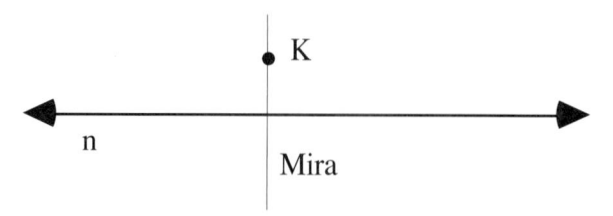

To use the Mira to obtain the construction in number 10, first draw a line through point Q perpendicular to line m, using the procedure shown above. Then turn the Mira so that it is approximately parallel to line m and again place one end on point Q. Then use the Mira in the same way to draw a line that is perpendicular to the new line just drawn. This line will be parallel to line m because two lines perpendicular to the same line must be parallel.

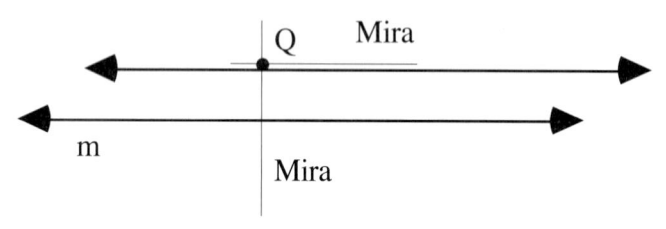

15. a. A circle circumscribed around the regular pentagon goes through all of the vertices of the pentagon. To find the center of the circle construct the perpendicular bisectors of two of the sides of the pentagon. Follow the procedure in Example J to construct perpendicular bisectors. The point where they intersect is the center of the circle. When the center is located, set a compass to the distance between the center and one vertex. This is the radius of the circumscribed circle.

    b. The regular hexagon can be circumscribed using the same procedure as in part a. An easier method is to use a straightedge to connect pairs of opposite vertices. These will also intersect in the center of the circle.

17. a. Here are two possible quadrilaterals whose sides are the given lengths.

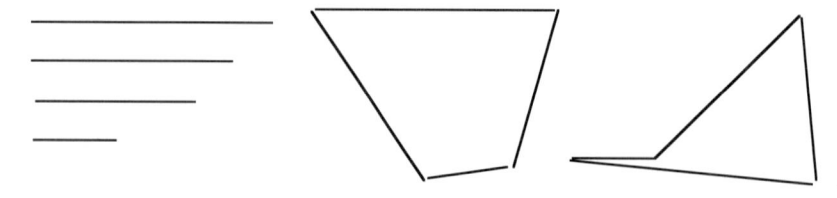

    b. If you imagine each vertex of the quadrilateral as being a hinged joint, so that the quadrilateral is formed by linkages, you can see that there are an infinite number of non-congruent quadrilaterals which can be constructed from the same four lengths. Imagine slightly changing one of the angles without changing any lengths of sides and observe what would happen to the other angles.

19. Drawings are shown below the descriptions of the constructions.

   a. Use a ruler to draw a 2 cm side. Then use the protractor to measure a 60° angle between the 2 cm side and the side of the triangle with unknown length. Temporarily extend this side with unknown length farther than it will need to go. With the ruler, measure a 3 cm length from the other end of the 2 cm side at the proper angle so that it intersects the other side at exactly 3 cm.

   b. Use a ruler to draw a 5 cm side. Use the protractor to measure a 45° angle between the 5 cm side and a second side of the triangle. Make this side also 5 cm long, either using the compass or the ruler. Join the other ends of these two sides for the third side of the triangle.

   c. Use a ruler to draw a 7 cm side. Since this is a leg and not the hypotenuse, use the compass or the protractor to construct a second side perpendicular to this one. Then construct a 35° angle either opposite or adjacent to the 7 cm side.

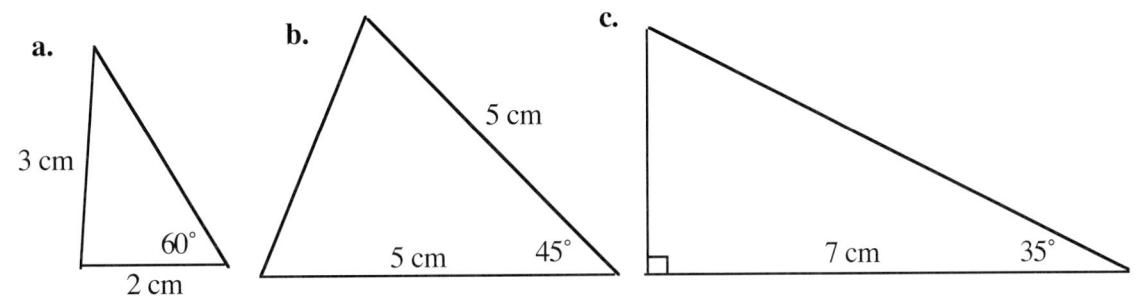

21. a. First use the compass to copy one length. Then construct the angle as in Example C. Finally, set the compass to the length of the second side, and mark it off so that the angle is included between the two given sides.

   b. For this construction, after copying one length and the angle, extend the length of the unknown side of the angle. Then set the compass to the second known length and mark the endpoints for the third side.

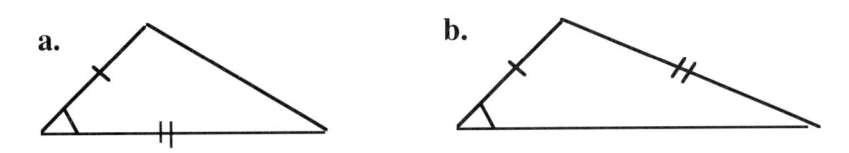

   c. The triangles are not congruent. The third side is longer in part b than in part a.

   d. We can conclude that having two pairs of sides and an angle congruent is not enough to guarantee the congruence of the triangles.

23. a. The diagram below shows that the given information is sufficient to conclude that △ABC ≅ △HMS. We have two pairs of sides and the included angle all congruent, so the triangles are congruent by the SAS congruence property.

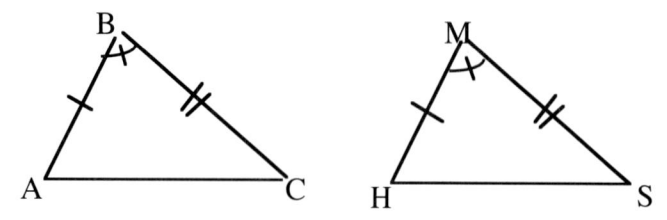

   b. This time the pair of congruent angles is not included between the pairs of congruent sides, so the triangles are not necessarily congruent. The diagram below shows why.

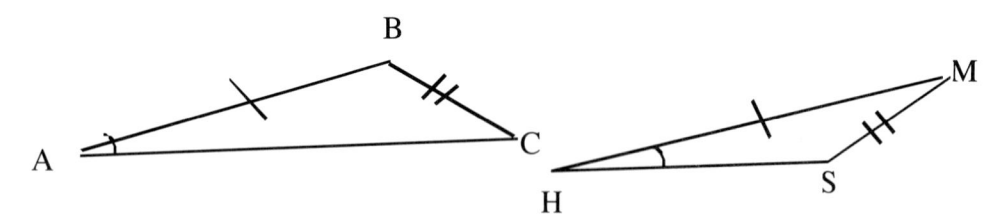

25. Construct AB and then with a compass open to span segment AB, swing arcs from both A and B. The point where the arcs intersect is the third vertex of the triangle.

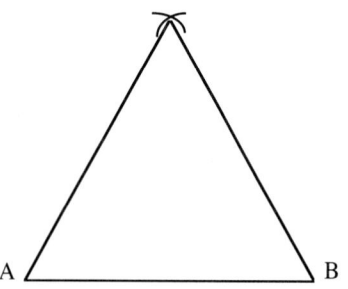

27. a. There are two possible points on line l for each distance. One is shown.

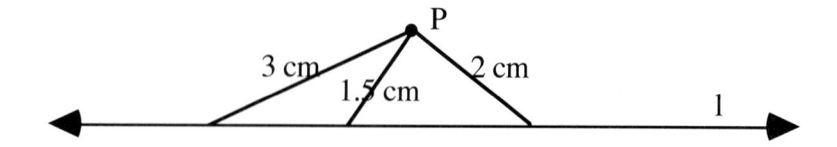

   b. The distance from point P to line l is about 1.1 cm.

   c. The shortest distance from a point to a line is the length of the perpendicular line segment from the line to the point.

29. a. The triangles are congruent by the SAS property. In each triangle a right angle is included between a pair of sides whose lengths match in the two triangles.

   b. They are congruent by the SSS property. The third congruent side is the one they share.

31. a. The triangles are congruent by the SAS property.

    b. The triangles are not necessarily congruent. The SAS property cannot be used because the congruent angles are not included between the congruent sides.

33. a. In the answer to #23b of this section it was possible to form two non-congruent triangles with the given angle and two given sides in the same relative positions of the two triangles. However, in this exercise it is not possible to do this. So, the triangles are congruent.

    b. The triangles are not necessarily congruent. We only know that one pair of angles and one pair of sides are congruent.

35. Since point K is the midpoint of segment RS, we know that RK ≅ KS. The triangles also share side KT and we are given that RT ≅ RS, so △ RKT ≅ △ SKT by the SSS property. Since the triangles are congruent we know that their corresponding parts are congruent. Therefore ∠R ≅ ∠S.

37. To trisect right ∠ABC, we will construct an equilateral triangle, so that we have a 60° angle, and then bisect this angle to form a 30° angle, trisecting the 90° angle. To do this, open a compass to span the distance BC and draw an arc with B as center. With the same compass opening, draw an arc with C as center and that intersects the first arc at point D. Then △BDC is an equilateral triangle and ∠DBC has a measure of 60°. Bisecting this angle will provide a trisection of ∠ABC.

39. Looking at the diagram we see that △ ABR ≅ △ CBD by the ASA property. This is true because ∠A and ∠C are congruent because they are both right angles, ∠ABR and ∠CBD are congruent because they are vertical angles, and side AB was measured to be equal to side BC. Side AR and side CD are corresponding parts of these congruent triangles, so their lengths are equal. So the distance across the river can be indirectly measured by measuring the equal distance CD.

41. The measured and unknown lengths shown form two congruent right triangles. Choose one of these triangles to look at, for example △ CBM on the right side of the gorge. Since this is a right triangle for which we know the lengths of two sides, we can use the Pythagorean Theorem to find the length of the third side. If let x represent the length of side MB, then we have $x^2 + 96^2 = 120^2$. Solving for x we get $x^2 = 120^2 - 96^2 = 5184$. So $x = \sqrt{5184} = 72$. Since AM = MB, the total distance from point A to point B is 72 + 72 = 144 feet.

## Section 11.2

1. a. No, the photographs are not quite congruent. It looks like they have been taken at different times, as the person on the platform has moved.

    b. The six stagings appear to be congruent.

    c. This picture suggests a translation mapping. The staging is shifted over to the right in each section of the photo.

3. The image in a translation appears exactly the same as the original image, just shifted to a different position.

   a. Since I is the image of F, the distance from F to I is the same as the distance from G to its image.

   b. The area of the original hexagon is exactly the same as the area of its image.

5. Here is a sketch of the original pentagon and its image from reflecting about line l.

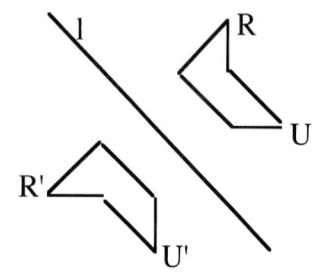

   a. The line through points R and R' will be perpendicular to line l, so the angles formed by these lines will have measure 90°.

   b. The distance from U to line l is equal to the distance from U' to line l.

   c. The fixed points for this mapping are all of the points on line l. These are the points whose locations are not changed by the reflection.

7. Below are the quadrilateral and its image after a 90° counterclockwise rotation about O.

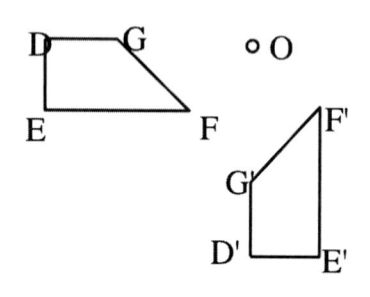

   a. The measure of ∠ FOF' is 90° because the rotation was through an angle of 90°.

   b. Lengths are not changed in a rotation, so DF = D'F'.

   c. Areas are not changed in a rotation, so the area of the original quadrilateral is equal to the area of its image.

9. a. The net effect of moving "right 4 and down 1" and then moving "right 2 and up 2" is the same as the single move "right 6 and up 1". This is the translation that maps A to A".

   b. If we compose the mapping that takes A to A' with the mapping that takes A' to A, then we will be moving the quadrilateral one place and then moving it right back along the same path. The image for the composition is just the original quadrilateral. This would be the identity mapping.

11. a. The composition of two reflections is always the same as a single translation. In this case it is a translation of 8 units horizontally to the left.

b. Since R is 7 units to the left of line n, R' is 7 units to the right of line n. This puts R' at a position 11 units to the right of line m, which means that R" is 11 units to the left of line m. So the distance from R to R" is 11 – 3 = 8 units. If you try this with some other points you will see that in each case the distance from a point to its image under the composition of the two reflections is always 8 units. This is double the distance from line m to line n.

13.

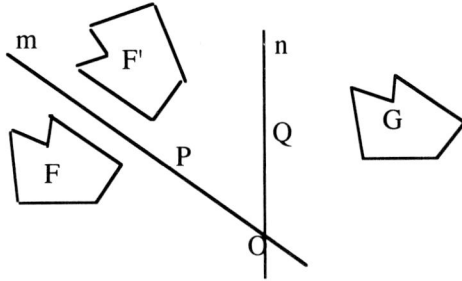

a. Call the image of F for a reflection about line m, F'. Then the image of F' for a reflection about line n is hexagon G.

b. We can also rotate hexagon F about point O to arrive at hexagon G. The number of degrees for this rotation is twice the number of degrees in angle POQ.

15. a. Draw KO and then construct a perpendicular to KO through point O. Use a compass to locate K' on the perpendicular so that KO = OK'. Then K' is the image of K. Obtain the images of the remaining 5 vertex points in a similar manner.

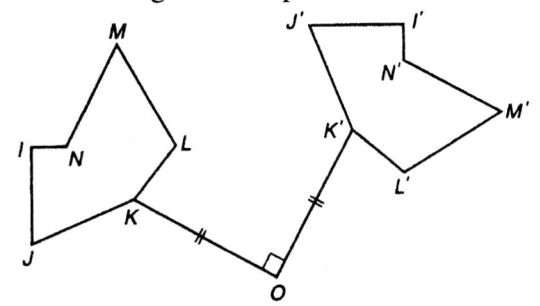

b. Construct the perpendicular through point Q to line k. Use a compass to locate Q' on the perpendicular so that the distance from Q' to line k equals the distance from Q to line k. Then Q' is the image of Q, and images of the remaining 4 vertices can be obtained in a similar manner.

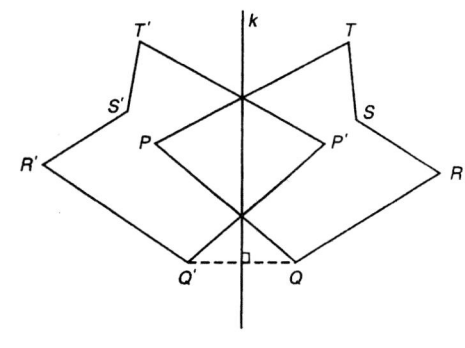

17. a. A 90° clockwise rotation about R brings the figure to a position so that its longest side is vertical and on the left side of the figure. Another 90° clockwise rotation about S brings it to the position shown in figure D below.

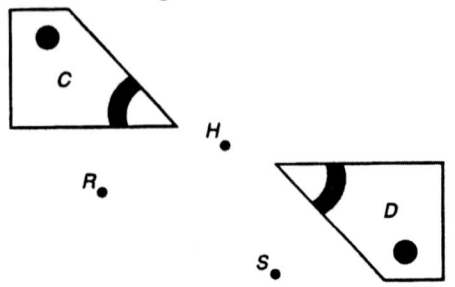

   b. Figure C can be mapped to figure D by a 180° rotation about point H.

19. For the row reflection, looking at the top row, squares 1, 3 and 5 will look identical and squares 2 and 4 will be the reverse. Reflecting about the lower sides of the squares will not create different figures, because there is a horizontal line of symmetry through the center of each square. The completed figure is shown at right.

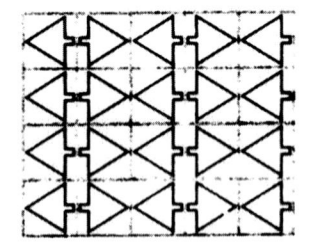

21. a. The center of rotation is on the tip of one arm of the figure. See the diagram below.

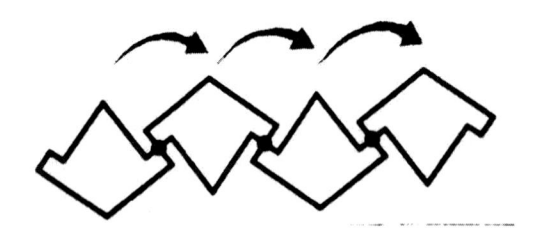

   b. To generate the left column of figures from the figure in the upper left corner, there are two options. Either reflect the figure over a horizontal line at its lower point, or rotate the figure 180° about the lower point.

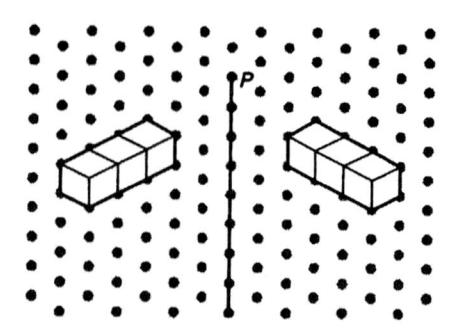

23.

## 11.2 Congruence Mappings

25. a. A rhombus has reflection symmetries through the two lines that go through opposite vertices of the rhombus. It also has rotation symmetries of 180° and 360°, so there are four symmetries.

    b. An equilateral triangle has three reflection symmetries, one line through each vertex and opposite midpoint, and rotation symmetries of 120°, 240°, and 360°. A total of six.

    c. A regular octagon has sixteen symmetries. There are 8 rotation symmetries, and 8 reflection symmetries, one through each pair of opposite vertices and one through each pair of midpoints of opposite sides.

27. The tessellating figure here was created by cutting pieces from three alternating sides of a regular hexagon and attaching them to the adjacent sides.

29. This tessellating figure was created by cutting a shape from the right side of a rectangle and attaching it to the left side.

31. There are many different reflection tessellations that can be made. For an example and the steps in the procedure, see Figure 11.34 in the text.

33. Create a curve from E to F and translate it to side GH. Then create a curve from F to H and translate it to side EG. This figure shows one example.

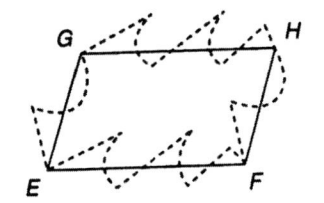

35. Create a curve from P to Q and rotate it about Q so that P maps to R. Similarly, create a curve from R to S and rotate it about S, and create a curve from T to U and rotate it about U. One example is shown here. Another example appears in exercise #27.

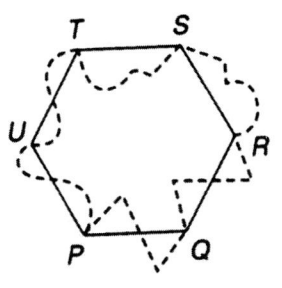

37. a. This mapping is a rotation. To find the center of rotation draw line segments connecting two of the vertices of the figure with their corresponding vertices in its image. Construct the perpendicular bisectors of these two segments and extend them until they intersect. This intersection point is the center of rotation.

    b. This mapping is a rotation. Draw segments joining A to A' and B to B' and find the intersection point of the perpendicular bisectors of these segments to locate the center.

39. a. Since the reflection is about the x-axis the x-coordinate of each point and its image will be the same. The y-coordinates are opposites. The coordinates of the image points are:
    D'(3, −5),    E'(1, −2),    F'(3, −3),    G'(5, −1).

b. This time the y-coordinates remain the same as for D'E'F'G' and the x-coordinates are opposite. So now both coordinates are the opposite of those in DEFG. The coordinates are: D"(⁻3, ⁻5), E"(⁻1, ⁻2), F"(⁻3, ⁻3), G"(⁻5, ⁻1).

c. A 180° rotation about the origin will produce the same effect as the two reflections.

41. Since the numbers that are placed in the three corners get counted twice in the side sums, we get different total sums when we place different numbers in the corners. For example, when we place 1, 2, and 3 in the corners then they each get counted twice and 4, 5, and 6 get counted once for a total of 27. This means we want each side to sum to 9. By placing different sets of numbers in the corners we can get different total sums. Those that are divisible by three give solutions to the puzzle. And if the sets of numbers in the corners are different then the figure can not be rotated or reflected onto itself. The four different solutions are shown below.

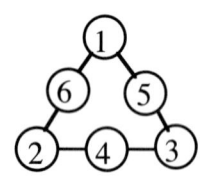

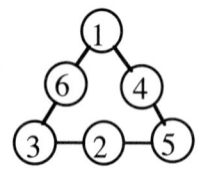

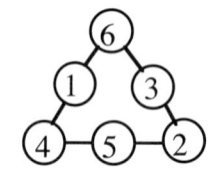

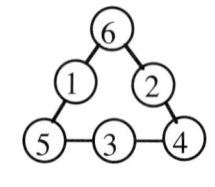

43. a. A translation of this frieze pattern will map it onto itself.

b. This frieze pattern can be mapped onto itself either by a translation or by a horizontal reflection through its center.

45. Rectangles, regular hexagons, and regular octagons are some examples of figures which can be mapped onto themselves by 180° rotation, translation, horizontal reflection, and vertical reflection. A frieze pattern based on variations of one of these would work. One example is shown below.

47. There are six additional magic squares that can be obtained from figure a, besides figures a and b. Figure b is from a reflection about one diagonal. We can also reflect about the other diagonal. We can reflect about the middle row or reflect about the middle column. So there are four different squares obtained by reflections. There are also four different squares obtained by rotations of 90°, 180°, 270°, and 360°.

49. a. There are three different ways to start the game. Either play in a corner, a middle of a side, or the center.

  b. A reflection about the upper left to lower right diagonal maps one grid onto the other. Below are shown the five possible noncongruent second moves.

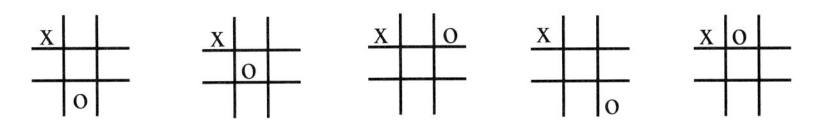

  c. Of the five moves shown above playing in the center is the only one that prevents an expert playing x from winning. *[Find the winning strategies for x in the other games.]*

## Section 11.3

1. a. Because the distance from the projection point O is twice as far to a point on the image as to the corresponding point on the figure, the scale factor is 2.

  b. Because the distance from the projection point O is half as far to a point on the image as to the corresponding point on the figure, the scale factor is ⁻1/2. It is negative because the original figure and its image are on opposite sides of the projection point. This turns the image upside-down.

3.

  Scale factor 2    Scale factor $-\frac{1}{3}$

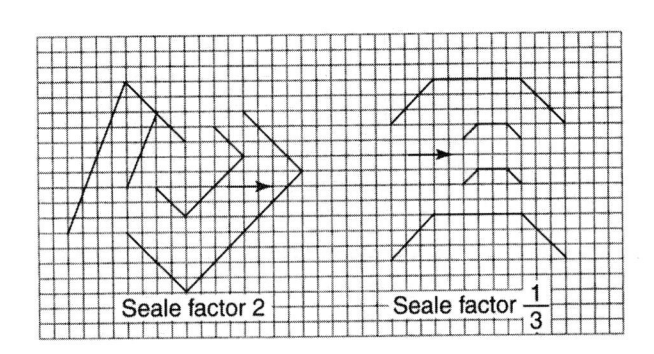

5. If point O is considered the origin for the coordinate grid, then projecting from O with a scale factor of 2 will take the point (x,y) to the point (2x,2y). In general, if the scale factor is k, then the image of the point (x,y) is the point (kx,ky).

7. a. Since the polygons are similar, the lengths of corresponding sides are proportional. The ratio for corresponding sides here is 2 to 3, because YZ = 2 and UV = 3. To find the length RV we set up the proportion $\frac{2}{3} = \frac{4}{RV}$. So RV = 6. Similarly, because $\frac{2}{3} = \frac{5}{SU}$, we have SU = 7.5, and $\frac{3}{2} = \frac{4.5}{WX}$, giving WX = 3.

  b. Here the ratio is 2.6 to 3.9, which is again equivalent to 2 to 3. Using proportions as in part a. we find that AD = 4.6, NP = 3, and NO = 6.6

9.  a. △ABC ~ △EFD because all of the corresponding angles are congruent. Since the sum of the angles in a triangle is 180°, the measure of angle C is 180 – 30 – 95 = 55°.

    b. △GHI ~ △JKI because they satisfy the angle-angle similarity property for triangles. ∠ GIH is congruent to ∠ JIK because they are vertical angles.

    c. These triangles are not necessarily similar. They are both isosceles, so ∠ M = ∠ N and ∠ R = ∠ S, but we don't have any angles of △LMN equal to any angles of △TRS.

11. a. Any two squares are similar because all of the angles are 90°, so all corresponding angles are congruent and pairs of corresponding sides will also all have the same ratio.

    b. No. The figure in #9c. shows an example of two isosceles triangles which are not similar.

    c. No. A pair of rhombuses that are not similar are shown below.

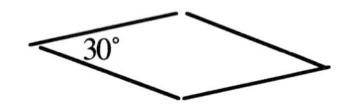

    d. Any two regular octagons are similar. All of the interior angles are always 135° and corresponding sides will have the same ratio because all of the sides of any given regular octagon are congruent.

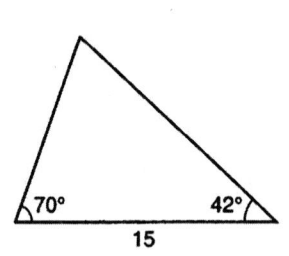

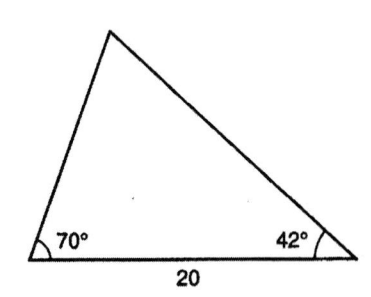

13. a.                                           b.

    c. The triangles shown above are similar. All corresponding angles are equal and lengths of corresponding sides are all in the ratio 3 to 4.

    d. This suggests that if two angles of one triangle are congruent to two angles of another triangle, then the two triangles are similar.

15. △ABC is similar to △AEF because they both contain angle A and a right angle. Therefore, AB/AE = BC/EF = AC/AF.
    Also, △ACD is similar to △AFG, so AD/AG = DC/GF = AC/AF.
    All angles in both rectangles equal 90°. Therefore, the corresponding sides of ABCD and AEFG are proportional and their corresponding angles are equal.

17. a. We are given that ∠ STR = ∠ CAB because they are both formed by the sun's rays and we also know that ∠ ABC = ∠ TRS because both are right angles, so △ABC ~ △TRS by the angle-angle similarity property.

   b. Since the triangles are similar, corresponding sides are proportional. Converting all measurements to meters, we get $\frac{2}{35} = \frac{0.8}{RS}$, so RS = 14. The tree is about 14 meters tall.

19. a. The scale factor of the figure to the image in part a of exercise 4 is 3 to 1. The area of the image is 9 times greater than the area of the figure because the square of the ratio 3 to 1 is 9 to 1.

   b. The scale factor of the figure to the image in part b of exercise 4 is 1 to 2. The area of the image is 1/4 of the area of the figure because the square of the ratio 1 to 2 is 1 to 4.

21. a. A mapping with a scale factor of 1/2 will reduce the area by a factor of 1/4.
   Since triangle T has an area of 16 sq. units, its image will have an area of 4 square units.

   b. A mapping with a scale factor of 3 will enlarge the area by a factor of 9.
   Since triangle T has an area of 16, its image will have an area of 144 square units.

23. We can count cubes to find that the volume of the larger figure is 56 cubic units and count squares on its surface to see that its surface area is 104 square units. Note that 104 = 4 × 26 and 56 = 8 × 7. For a scale factor of 3 the surface area will be $3^2$ = 9 times the surface area of the small figure and the volume will be $3^3$ = 27 times the volume of the small figure.
   In general, the surface area increases by the square of the scale factor and the volume by the cube of the scale factor. The completed table is shown below.

   | Scale Factor | Surface Area (square units) | Volume (cubic units) |
   | --- | --- | --- |
   | 1 | 26 | 7 |
   | 2 | 104 | 56 |
   | 3 | 234 | 189 |
   | 4 | 416 | 448 |
   | 5 | 650 | 875 |

25. The height of the enlargement is 3 times the height of the original, so the scale factor is 3.

226                    Chapter 11  Motions in Geometry

27. The scale factor for this enlargement is 2 because each point in the enlargement is twice the distance from the origin as the original point.

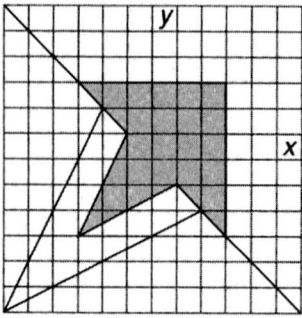

29. For this mapping the figure and its image are on opposite sides of the projection point, so the image is turned "upside-down" and is enlarged by a factor of 3, so the scale factor is ⁻3.

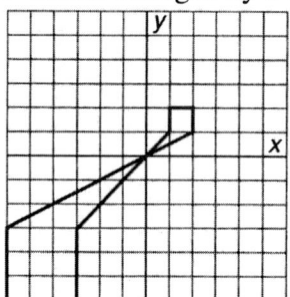

31. a. Since Beth's friend's height and the length of her shadow were the same, the height of any vertical object would be the same as its shadow's length at that time of day. The triangles formed by any two such objects, their shadows, and the sun's rays would be similar triangles (in this case they would be isosceles right triangles). So the height of the goal posts was 10 yards, or 30 feet.

   b. If the posts had been 10 feet higher, then their shadow would have been 10 feet longer, 40 feet instead of 30 feet, or $13\frac{1}{3}$ yards instead of 10 yards.

33. a. Area varies as the square of the scale factor, so the surface area of the wings of the full-size plane is $15^2 = 225$ times as much as the surface area of the wings of the model.

   b. Because volume varies as the cube of the scale factor, the volume of the full-size plane is $15^3 = 3375$ times as much as the volume of the model.

   c. Distance is a one-dimensional measure. The tip of the wing of the full-size plane flaps a distance of $15 \times 2$ cm = 30 centimeters.

35. Sketching a few images of the pentagon, all with a scale factor of two, should convince you that no matter where the projection point is placed, all of the pentagons will be congruent. They will all have sides twice as long as the original and areas four times as large.

37. a. The original sheet is similar to the quarto with a scale factor of 2 to 1, because the two adjacent edges of the original sheet were both folded in half to obtain the quarto; and the folio is similar to the octavo with a scale factor of 2 to 1 because two adjacent edges of the folio were both folded in half to obtain the octavo.

b. After every even numbered fold a rectangle similar to the original is obtained, and after every odd numbered fold a rectangle similar to the folio is obtained.

39. If the rubber bands maintain the ratio of the distance from O to P' being 2.3 times the distance from O to P, then the enlargement will have a scale factor of 2.3 to 1. Since distance is a one-dimensional measure, the distance from Denver to Kansas City on the enlargement will be 2.3 times greater than the distance on the small map.

# Chapter 11 Test

1. a. Using any compass opening and point A as center, draw arcs intersecting the sides of ∠A in points B and C. Then draw arcs with points B and C as centers using the same compass opening and so that the two arcs intersect. Call this point of intersection D. Then the ray from A through D is the angle bisector.

b. Using the same compass opening and point Q as center, locate points A and B on line m so that AQ = BQ. With the same compass opening and with A and B as centers, draw arcs intersecting at point D in one half-plane and point C in the other half-plane. Then line segment DC is perpendicular to line m.

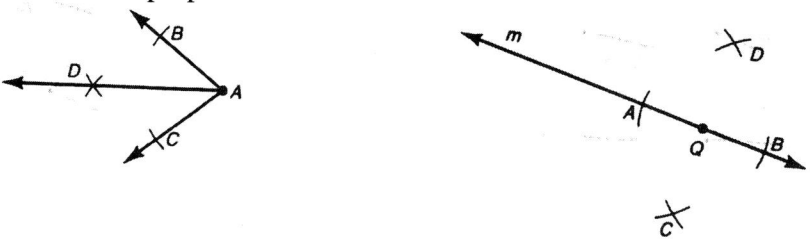

c. Using the same compass opening throughout, draw arcs so that PA = AB = BC = PC, as shown in the figure below. Then quadrilateral PABC is a rhombus and the line through points P and C is parallel to line l.

d. With R as center, draw arcs intersecting line n so that RA = RB. With the same compass opening and points A and B as centers, draw arcs intersecting at D so that AD = BD. Then the line through points R and D is perpendicular to line n.

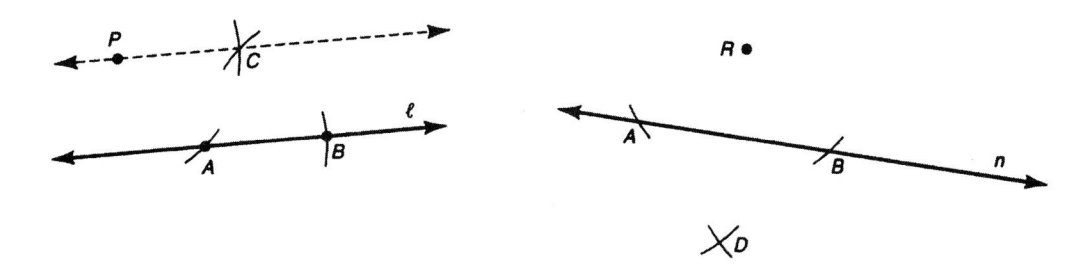

2. The center of the circumscribed circle is located at the point where the perpendicular bisectors of the sides meet. Choose two of the sides and construct their perpendicular bisectors. For example, construct the perpendicular bisector of AB by using the same compass openings at points A and B and drawing arcs on both sides of AB. Join the intersection points of the two arcs to construct the perpendicular bisector. Place the compass point at the intersection point of the perpendicular bisectors and open it to meet one of the vertices of the triangle. Then draw the circumscribed circle.

3. a. Use a compass to copy the length of each segment and join the endpoints of segments to form the triangle.

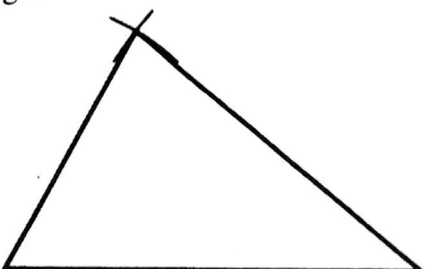

   b. It is not possible to construct a triangle with these three lengths because one of the sides is longer than the sum of the other two sides.

4. a. ∠ACB ≅ ∠DCE because they are vertical angles. So △ ACB ≅ △ DCE by the ASA congruence property.

   b. Side FI is shared by both triangles and ∠GIF ≅ HIF because they are both right angles. So △ GIF ≅ △ HIF by the SAS congruence property.

   c. These triangles are not necessarily congruent. The pair of congruent angles is not included between the pairs of congruent sides.

   d. There are three pairs of congruent sides since side RT is shared by both triangles. So △ QRT ≅ △ STR by the SSS congruence property.

5. a. Imagine sliding the figure down two units and three units to the right. Map each vertex to the corresponding point that is down two units and three units to the right.

   b. Draw a perpendicular line from each vertex to the point that is the same distance on the other side of line l. Then connect these new vertices to form the reflection.

   c. For each vertex of the original figure, connect it to point O and then form a 90° angle with the other side of the angle the same length and 90° in the clockwise direction. The endpoint of the new segment is a vertex of the image. Connect the new vertices to form the image.

6. a. If the center of rotation is the same, then the composition of two rotations will be a single rotation. In this case we move clockwise 45° and then back counterclockwise 70° so the composition is a counterclockwise rotation of 25°.

   b. As we saw in Sec. 12.2 exercise #11, the composition of the two reflections will be a translation. It will translate points twice the distance between lines l and m.

   c. A composition of two translations is another translation. If we move on a grid right 12 units, then up 8 units, then right 5 units, then down 10 units, we can accomplish the same movement by moving right 17 units and down 2 units.

7. The number of mappings of a plane figure onto itself is the total number of lines of symmetry and rotation symmetries.

   a. An isosceles triangle has two equal sides and two equal angles. This gives it one line of symmetry. It also has one rotation symmetry, through 360 degrees. So it has two congruence mappings.

   b. A regular hexagon has six lines of symmetry and six rotation symmetries, for a total of 12 congruence mappings.

   c. A rectangle has two lines of symmetry and two rotation symmetries, for a total of 4 congruence mappings.

8. a. To create an Escher-type tessellation starting with an equilateral triangle, we can follow the procedure for rotation tessellations as outlined in section 11.2 of the text.

   b. To create an Escher-type tessellation starting with a parallelogram, we can follow the procedure for translation tessellations as outlined in section 11.2 of the text.

9. a. We can translate the frieze any number of units to the left or right, where a unit is the length of the repeating segment. We can also reflect around a vertical line that goes through any of the peaks or valleys of the large sawtooth part of the pattern.

   b. Again we can translate any number of units to the left or right. In this one we can also rotate through 180 degrees.

   c. Translations again work here, but rotations will not. However, we can perform a glide reflection by sliding left or right the distance between two of the small triangles, and then reflecting across a horizontal line through the center of the frieze.

   d. This one has many possibilities. As before, we can translate left or right. We can rotate 180 degrees. We can reflect across a horizontal line through the center or reflect across a vertical line either through the peak or valley of the sawtooth. And a glide reflection in which we translate between the small central figures before reflecting would also work.

10. a. Draw a segment from point O to one of the vertices of the figure. Extend this segment so that its total length is twice the distance from O to the vertex. At the end of this segment place the point that is the image of this vertex. Repeat with each of the other two vertices and connect them to form the image.

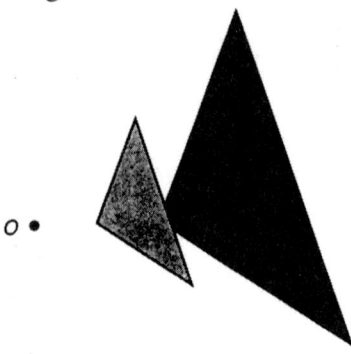

   b. Draw segments as in part a., but this time place the image points 1/3 of the distance from O to the points in the figure.

   c. Since the scale factor is negative, the image points will lie on the opposite side of O from the original figure, and 1/2 the distance from O.

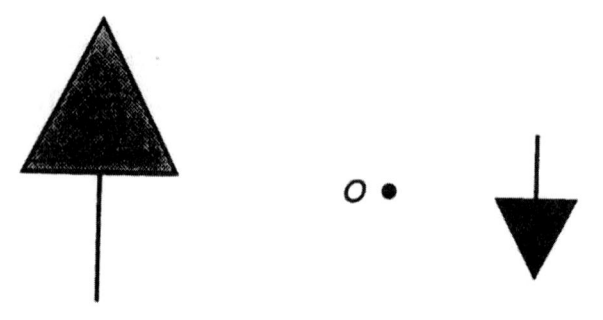

11. a. Since ∠BCD ≅ ∠ACE and ∠BDC ≅ ∠AEC,
      △ BCD ~ △ ACE by the AA similarity property.

   b. ∠FHG ≅ ∠IHJ because they are vertical angles and ∠GFH ≅ ∠JIH,
      so △ FHG ~ △ IHJ by the AA similarity property.

   c. These two triangles are not necessarily similar. We only know that they have one pair of congruent angles, and we only know the ratio of one pair of sides.

   d. △ RST ~ △ WUV by the SSS similarity property because the lengths of all three pairs of corresponding sides are in a ratio of 2 to 1.

12. a. Two rectangles are not necessarily similar. One counterexample is shown below.

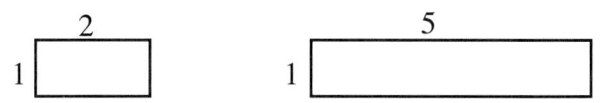

   b. Two squares are always similar. All of the angles are 90° and since all four sides of a square are the same length, lengths of all pairs of corresponding sides of any two squares will have the same ratio.

   c. Two right triangles are not necessarily similar. One counterexample is shown below.

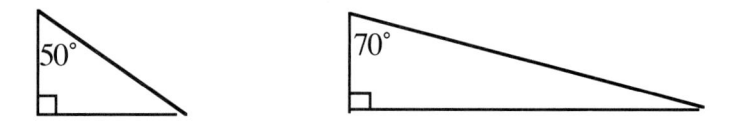

   d. Two equilateral triangles are always similar. They would satisfy the AA similarity property for triangles because all of the angles are 60°.

   e. Any two congruent figures are also similar. All pairs of corresponding sides have lengths in the ratio of 1 to 1, and all corresponding angles are congruent.

13. a. The dimensions of the original figure are 3 by 2 by 2. An enlargement with a scale factor of three has dimensions 9 by 6 by 6, so the volume is 9(6)(6) = 324 cubic units. We can also get the answer by multiplying 12 by $3^3$ because the volume will increase by the cube of the scale factor.

   b. Area will increase by the square of the scale factor, so the surface area of the enlargement will be 32 × 9 = 288 square units.

   c. When we use a scale factor of 1/2, we reduce the volume by a factor of $(1/2)^3 = 1/8$. The volume of the reduction is 12 ÷ 8 = 1.5 cubic units.

   d. When we use a scale factor of 1/2, we reduce the area by a factor of $(1/2)^2 = 1/4$. The surface area of the reduction is 32 ÷ 4 = 8 square units.

14. a. Height is a one-dimensional measure, so the height of the model is 1/4 the height of the table. The model's height is 28 ÷ 4 = 7 inches.

   b. The amount of stain required depends on the surface area. Area is a two-dimensional measure, so the actual table's surface area is $4^2$ = 16 times the surface area of the model. The table requires 16 times as much stain, which is 16 fl. oz. or 1/2 quart.

   c. The weight corresponds to the volume, which is a three-dimensional measure. The actual table's weight is $4^3$ = 64 times the weight of the model. The weight of the table is $\frac{3}{4} \times 64 = 48$ pounds.

15. The person, the beam of light and the person's shadow form a right triangle which is similar to the triangle formed by the light pole, the beam of light and the shadow of the pole. Since the triangles are similar, their corresponding sides are in proportion. If we call the height of the streetlight x, then $\frac{6}{10} = \frac{x}{45}$. So 10x = 6(45) = 270 and x = 27. The streetlight is 27 feet tall.

16. a. Triangles AIB and CDB are not necessarily congruent. Unless the campers happen to walk the same distance from B to C as they did from A to B, the triangles will not be congruent. The triangles are similar. Angles IAB and DCB are congruent because they are both right angles. Angles ABI and CBD are congruent because they are vertical angles. Since the triangles have two pairs of congruent angles, the third pair must also be congruent, and the triangles are similar.

   b. Because corresponding sides of similar triangles are proportional, we can set up and solve a proportion to find the missing length. There are various correct ways to set this up. Here is one. The ratio of AB to AI is the same as the ratio of CB to CD. So if we call the length from A to I x, then 300/x = 24/40. Then 24x = 300(40) = 12000. So the distance to the island is 12000/24 = 500 feet.